ADVANCED COMBAT
HELICOPTERS

Edited by Mi Seitelman - Design by David Polewski

Motorbooks International
Publishers & Wholesalers Inc
Osceola, Wisconsin 54020, USA

Printed in Japan

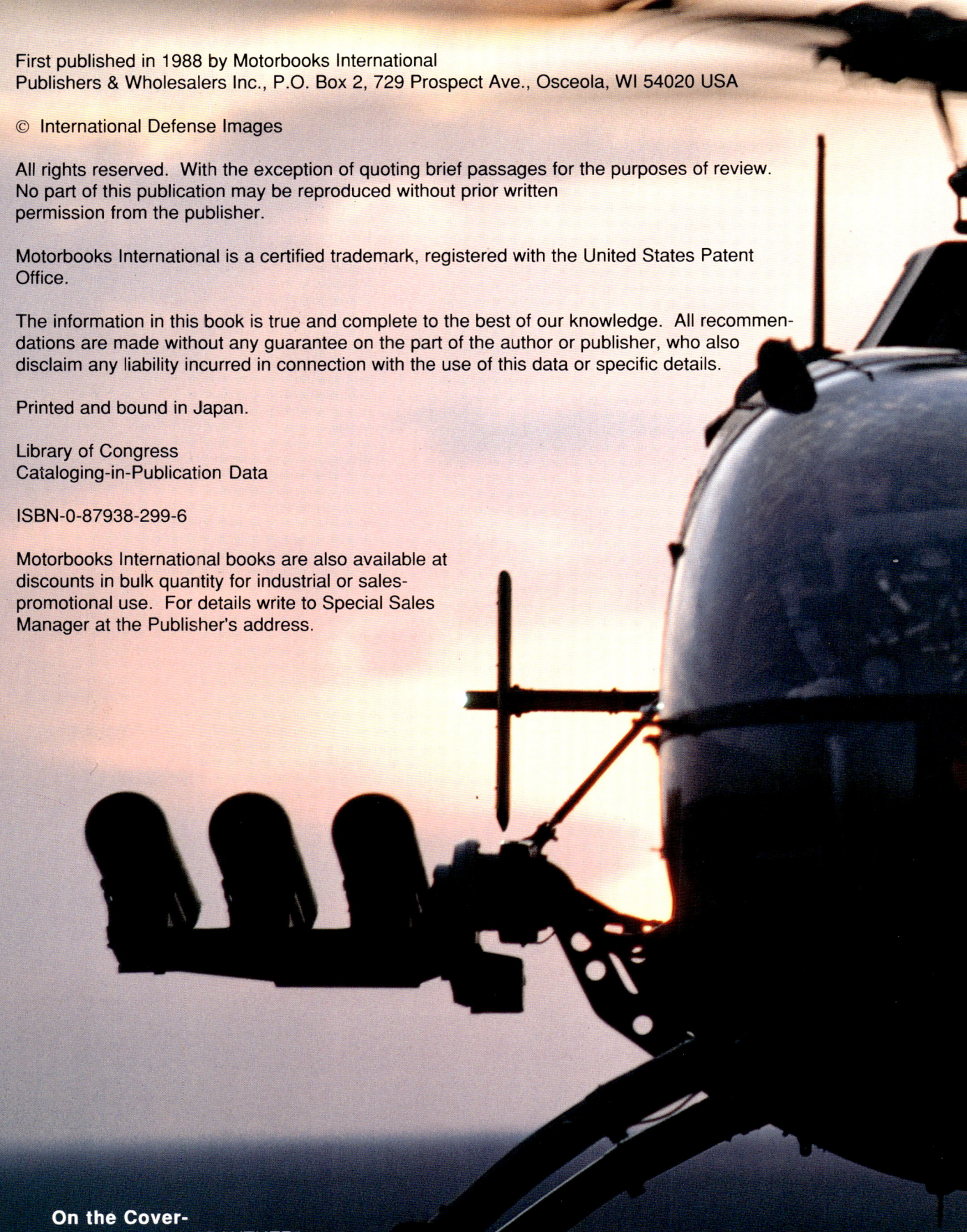

First published in 1988 by Motorbooks International
Publishers & Wholesalers Inc., P.O. Box 2, 729 Prospect Ave., Osceola, WI 54020 USA

© International Defense Images

All rights reserved. With the exception of quoting brief passages for the purposes of review. No part of this publication may be reproduced without prior written permission from the publisher.

Motorbooks International is a certified trademark, registered with the United States Patent Office.

The information in this book is true and complete to the best of our knowledge. All recommendations are made without any guarantee on the part of the author or publisher, who also disclaim any liability incurred in connection with the use of this data or specific details.

Printed and bound in Japan.

Library of Congress
Cataloging-in-Publication Data

ISBN-0-87938-299-6

Motorbooks International books are also available at discounts in bulk quantity for industrial or sales-promotional use. For details write to Special Sales Manager at the Publisher's address.

On the Cover-
The Aerospatiale "PANTHER" light tactical transport and combat helicopter.
Photo by Frederick Sutter

Contents

Profile: L.M. "Jack" Horner
Page 4

AH-1W SuperCobra
Page 8

Air-to-Air Combat
Page 10

CH-53D/E
Page 13

Apache AH-64
Page 17

Chain Gun
Page 33

OH-58D
Page 37

SH-60B
Page 42

NATO'S Combat Helicopters
Page 46

LYNX
Page 46

Panther
Page 53

BO 105/PAH-1
Page 65

Mongoose
Page 73

AH-1S Cobra
Page 81

CH-46/CH-47
Page 84

BO 105/PAH-1

--- EDITORIAL STAFF ---
SUSAN MALINOWSKI-TURNER • PHILIP FARRIS • GARY KIEFFER

PHOTOGRAPHY	
FREDERICK SUTTER	Cover 1, Page 1, 2-3, 5, 7,9 (lower),17, 20-21, 22, 23, 26, 27, 28, 29, 30, 31, 38, 39, 40-41, 53 thru 80, 84-85
MI SEITELMAN	Cover 4, Page 8-9, 11 (bottom), 13, 14-15, 16,36, 42-43 (bottom), 43 (center), 44, 45, 86-87, 94-95
GEORGE HALL	Page 22 (right-center), 26-27 (bottom), 33
ROBIN ADSHEAD	Page 49, 50, 51, 52
BRIAN R. WOLFF	Page 18-19, 28-29 (top), 32
DAVID HATHCOX	Page 42 (top), 43 (top), 90, 91
GARY KIEFFER	Page 88-89, 92-93 DAVID POLEWSKI Page 96

JONATHAN SCOTT ARMS   Stock Photography/Research

PROFILE: Leonard M. "Jack" Horner
President, Bell Helicopter Textron, Inc.

J. Philip Geddes

Bell Helicopter, the world's largest helicopter manufacturer, was founded in 1935 by Lawrence D. Bell as an aircraft company. Work on Bell's first helicopter began in 1941 near Buffalo, N.Y., where their first machine flew in 1943. By 1987, Bell had produced over 27,000 helicopters, more than all other manufacturers in the world put together. Always a massive force in the design and production of both civil and military helicopters, Bell is now building four different commercial types at its Fort Worth, Texas, plant, and two in Canada. Civil ouput ranges in size from five- to 20-passenger capacity. Bell currently has five military helicopters in its product lines for the U.S. Army, Navy and Marine Corps. Production at the present time is greatly reduced from the levels of the late 1970s.

The future of vertical lift for the next several decades is being shaped by two major military programs, the V-22 Osprey tilt rotor and the Army's LHX (Light Helicopter Experimental). Bell originated the tilt rotor concept with the NASA/Army/Navy XV-15 research aircraft. This pioneering proof of concept led to a $1.714 billion fixed price incentive contract, awarded on May 2, 1986, to Bell in association with Boeing Vertol (now Boeing Helicopter) Company for a seven year full scale development program. The V-22 is a joint Navy/Marine Corps/Air Force effort, in a plan to develop and produce over 1,200 units for a variety of missions, starting with amphibious assault for the Marines in 1991. LHX is a paper helicopter of changing form, being hotly pursued by all of the heavyweights in the U.S. helicopter industry. Bell, in this case, is teamed with McDonnell Douglas Helicopters. The prize is enormous, with the Army planning to buy approximately 4,500 aircraft that will replace several existing scout, attack and utility helicopters, beginning in the 1990s. The V-22 and LHX will revolutionize the effectiveness of vertical lift — the first through its unique flying abilities and the other with future vision electronics.

Mr. Horner, is a hard driving, congenial executive, known to speak his mind. A native of West Hartford, Connecticut, he graduated from Yale University with a B.S. in Industrial Administration. Mr. Horner saw war at first hand in Korea, winning four Air Medals as a Marine infantry officer and aerial observer. He learned to fly fixed-wing aircraft and helicopters before leaving the military. Mr. Horner joined Bell as Vice President of Operations in 1974 from Sikorsky. He was appointed President of the Textron subsidiary in 1983.
Mr. Horner's views were revealed in a free-wheeling interview with International Defense Images.

HORNER'S VIEWS

The entire success of this country has rested on people developing new products. Damage that innovativeness and you damage the country. My greatest concern is that we are not spending the right amount of research and development (R & D) dollars, even though the whole thrust of the economy ultimately revolves around those dollars. As the Department of Defense or the Congress squeeze profits, and major investments of cash go into funding of inventories, where interest is not an allowable expense, the payoff is further and further down the line. This has a detrimental effect on all of our R & D because most aerospace companies are in both the commercial and government business, and research flows both ways.

Bell is in a very fortunate position, with years of development on the tilt rotor, and other rotors and transmissions that are now paying off. My primary concern focuses on what happens down the line. In the late 1970s and early 1980s, when the commercial, non-U.S. government sector took off, we greatly benefitted and were able to invest between $60 million and $70 million a year in R & D. Back then, we were building 700 plus civil airplanes a year; U.S. government business was down to about 20 percent of the total. Now we are in a situation I would like to maintain, with the U.S. government representing about 50 percent of business. The V-22 Osprey is a major funded development program that in the long term has big payoffs, but not without great risks. This is because once again we are committed to a competitive environment that we still have to work our way through.

In the aerospace business you should be spending between four to six percent a year of sales volume on independent research and development (IR & D). Right now we are on the four percent side, although we were at six percent. Osprey's development (government funded) is very large, and beyond our IR & D.

INDUSTRY/GOVERNMENT RELATIONS

There is a perception, raised since the infamous days of the $400 hammer, that the defense business overcharges the customer across the board for any product. First of all, the defense business cannot be directly compared with commercially competitive types of business. Defense goods are only there to deter. They really don't have a positive, beneficial level other than in the technology that spins off into the other avenues, which has a tremendous value. In non-U.S. government business there is a multitude of markets, globally, for our product. When you deal with the Department of Defense, in most cases you have one customer, and one product for which there may not be any other markets. On top of that, you have an uncertain annual funding level through the Congress. Investment on this side of the marketplace creates an entirely different risk. Over the past four to five years a feeling has built up that defense contractors make too much profit and abuse the system. The variety of laws and regulations that have been added to the system ultimately reduce funds for what I classify as the critical part, the R & D portion of the business.

INDUSTRY/CONGRESSIONAL RELATIONS

Corporations have a difficult time communicating. Historically, we don't do a good job conveying a message of the good things we do for the community. The same things apply to discussions with Congress. Congressmen on an individual basis see what we are doing innovatively in manufacturing and in developing new products. They have some sympathy for the need to spend on R & D and other items on line, but our inability to get the beneficial side across to the total populace causes Congress to react to their constituencies. They get elected by people and many of the people who talk to them are negative about defense issues.

When I was president of the AUSA (Association of the U.S. Army), I would stand up at the end of large membership meetings and

ask everyone to go home and just tell one or two friends about what they heard, because most of the time we are all talking to ourselves. Fort Worth happens to be an exceptionally strong military town, with a large Air Force base nearby, that believes in the whole concept of a strong deterrent and defense industry. But we need to reach people who are not in that world. That is the only way negative perceptions are going to get solved in the long run. An initiative has already been taken by the major defense companies to publicly audit themselves and take visible actions to insure that the people at large have a better understanding of what is wrongly considered waste or fraud.

We have created a written credo in Bell that defines the way we think Bell ought to run. The customer is number one in the credo, but, not only that, you have relations with your vendors and relations with your employees, relations with the community, and the stockholders. People are going to make mistakes, no question about it. Anyone who does something illegal deserves to get fired.

MILITARY MUST PUSH STATE OF THE ART

The military really has to decide what is needed to defeat the threat. Industry needs to provide technology and what they think they can build for the services to meet that threat. The military has to push the state of the art to always be that much better than any enemy. Now when you are always pushing the state of the art, you are going to have failures. If, however, as we are seeing right now, every failure is looked upon as bad, something is wrong. The critics say the program wasn't managed right or the system wasn't designed properly, or some people weren't doing their jobs right. My point is that when you're pushing the state of the art, if you back away from innovation and everything becomes success oriented, pretty soon you don't stay first and once again you are focused into a single customer with a single set of requirements. In the commercial world, you are always trying to look at a broad use for an aircraft in several markets. And in the commercial world of helicopters, users want to push the state of the art in terms of reliable, low cost operation. The military also wants low cost of operation and high reliability, but they have got to push the state of the art for the best possible performance —further than you would push the civil side. In military operations, nobody comes in second. As important and self evident as all that may be, many people lose sight of those differences in the environment industry is in right now.

THE BRIGHT SIDE

In some cases the system works extremely well. The OH-58D program is a classic example of a well thought out specification, worked out between the contractors and the Army, that became a superbly managed program because everybody became part of the problem and worked towards a solution. We ended up with a good system and a good airplane that met schedule.

The AH-1W program is another — three years from fixed price contract to delivery. Also, a very good program from the point of view of the customer, but it cost us a little more to build than we anticipated as a result of our underestimating the cost of installing the systems in the first production airplanes.

In the V-22 Osprey program, the specifications were worked very well and the requirements well established, but we have a long way to go yet to see if a fixed price basis was the right way to contract for such an ambitious development effort. We believe we can make it. The real test will come in God's wind tunnel, when we get in the air and fly. It's possible that we will run into unknown situations that could cost more than anticipated. There is always a question in development on whether fixed price constrains the customer also. One of the major advantages of a cost type contract is that it allows the customer to keep a high input. Under a fixed price, the customer in essence becomes blocked out because in a lot of cases you have to pretty much adhere to the written word. Once again, you are trying to develop a system on an R & D basis that's going to provide what the customer wants to win a war. I believe 100 percent in a fixed price environment for production. Absolutely. The question remains as we go through an evolutionary cycle, of whether we will end up with what the customer really wants. I hope it will work. I really do.

V-22 A DREAM MACHINE

For the V-22 Osprey there is a multiplicity of mission scenarios from the three or four originally established that drove the design. Derivatives of the original V-22 can go on forever — as far as your imagination will take you: electronic warfare, ASW, carrier delivery, refueling, over- the-horizon targeting, are all there. The Osprey has the unique capability to operate in an environment where it is not a helicopter and not an airplane but does both more efficiently than any other VTOL machine ever built. The V-22 has so many plusses that a lot of people keep inventing missions. It starts out looking a lot like the original helicopter, without many missions. Now look what they do: logging, off-shore surveying, news gathering, executive transport and all the military uses. I don't think pioneers like Sikorsky, Bell, Piasecki and others ever envisioned what vertical takeoff and landing aircraft could do. And it's a young industry, only 40 plus years old.

SERVICES SHAPE NEW VTOLS

The Army is extremely interested in vertical takeoff and lift and makes it a major part of their whole weapon system. Their rationale on how they are going to beat the threat means they want better helicopters. Because of the shortage of people anticipated in the 1990s, they want helicopters that require fewer people to fly and maintain them. The Air Force, of course, is not a helicopter developer; they are helicopter users. They, too, see uses for VTOL machines in support activities and support the V-22 for a special mission type category.

Now when you get over into the affordability issue, with the constraints on defense budget spending, you never quite know where you're going to fall in terms of other priorities. Fortunately, there is strong support for the V-22 concept in the Congress because they also see commercial attributes for the aircraft and basically believe that the technology makes sense, but it has to make sense in the primary military missions you are looking at. The Marine Corps believes that it's a strong adjunct to their amphibious warfare concept. Not to be forgotten are the many good applications for just pure helicopters and there may be applications down the road for other types of VTOL aircraft we haven't thought of.

LHX — AN EVOLVING PROGRAM

The evolution of this requirement to defeat a particular threat has taken longer than usual. Part of the process involved seeking technology that maybe wasn't quite available. The basic rationale, which I happen to believe very strongly, is that in the next decade we are going to have fewer people available in both industry and the military. Secondly, we really have to work on education to make sure that we get better trained people. The Army's desire to have a one-cockpit, lower maintenance helicopter makes a lot of sense, but the evolving technology to do that is pushing the state of the art a bit. That kind of evolution takes a little longer to put together. LAMPS was along that line: it took time to evolve, but when it came out it was a super airplane. So there is a little bit of a genesis problem going on with LHX that may not be at all bad. The size of the relatively high speed vehicle, mission requirements, and lots of things are evolving, but there is no question of the need. Whether it is to arrive in 1996 or 1998 is not the driver; the driving force has to be the technology that the Army wants in the platform and the demonstrations of that technology and the mission equipment package.

Considering the helicopter as a platform, in that context, the mission equipment package is the driver. Whether it is going to be a one-man or two-man machine will make a big difference in size and weight. The engine was to be 1,200 horsepower (hp), designed to give a margin for growth around a light helicopter. Well the light helicopter grew heavier because the requirement grew, but the light one could come back. We don't know if the current sized LHX is what is truly required for the mission the Army wants to perform. If it isn't, you just have to use another engine, that's all. If that isn't right, maybe the size of the helicopter can be driven down? But I don't set requirements. We can play with tradeoffs. We will come up with vehicles that fit anything the customer wants.

The Air Force's Advanced Tactical Fighter demonstration/evaluation program, with two teams of people competing before going into production, may be the way LHX will finally go. There is a precedent right there that could be utilized. I believe that LHX should be built. I support the program and really think the technology is there to build a one-man airplane that can fight. I'm not sure we should build a one-man airplane that can be a fighter and a controller because there is just too much to be done in the cockpit.

When you look at the evolving technologies, from artificial intelligence and voice communications to thermoplastics, you are talking about seven or eight years from now and you have to say why not? You have to put all those things together in a box, fly them and make them work. LHX has all the bells and whistles, and one guy. This has all been proved on the F-18.

COMPETITION

Worldwide, you've got eight helicopter companies and three or four countries getting into the same business through offsets, co-production programs or technology transfer programs. There is a question whether the market can support all of them, so you are seeing a lot of teaming or cooperative ventures. There is a danger that when you crossbreed too much technology you don't get a lot of differences and new and better devices coming out in the future. If you get everybody working in the same vein, you end up with mediocrity. We need different ideas — Sikorsky with their X-wing, Bell with the tilt rotor, and so on. There are benefits from an engineering point of view in some instances in cooperation and detriments if you go too far.

Once you decide to license one technology, you had better have another one that's better. In some instances you are forced into licenses by the need to generate a business base now that will create R & D funds that will in turn create a better future market. In that case, 50 percent of something today is better than nothing. There is a question right now on whether we are creating a technology base in the Far East, for example, that over a long period of time could compete with us. Teaming, coproduction, licensing, all dilute the marketplace because they distribute certain manufacturing processes. My belief is that it's much like a patent which is only good until someone invents a better mechanism to defeat the patent. The technology of computer controlled machine tools, which in essence was an Air Force generated concept, went overseas and now the machine tool industry is fighting tooth and nail for survival. Long term effects are sometimes hard to determine.

Aerospatiale is our number one competitor. They have a product line that goes across ours and have done a good job in selling the world, no question about that. The other competitors are niches within Bell. For the next generation of machines, cooperation may be the only way to go because of high R & D costs. You see that with the big airplane manufacturers working together in consortiums in Europe with Airbus. Boeing is working with the Japanese, as is Bell. We have signed an agreement on tilt rotors with British Aerospace to take a look at the market there. We have to recognize that in these kinds of technologies and manufacturing processes, customers around the world are not only demanding, but getting, manufacturing in-country — man hours for expenses. If they buy something they expect you to offset. I think that is the way it's going to be and I don't think there is anything you can do about it except get with the right people.

AH-1W SuperCobra

The Bell AH-1W SuperCobra is the most recent, most powerful, and heaviest helicopter in the Cobra family. Developed to fill a U.S. Marine Corps' (USMC) requirement for an aerial platform that could offer hot-and-high performance, the AH-1W is equipped with an array of weapons that can include: rockets; air-to-ground missiles; a 20mm cannon; and counter-air-attack missiles.

The primary mission of the SuperCobra is to serve as an escort vehicle for the Marines troop-carrying helicopters, but it is also equipped to attack tanks and other ground vehicles, to provide close-in fire support, and to serve in reconnaissance and target-acquisition roles.

Improvements in the AH-1W go far beyond greater horsepower and firepower. A new head up display (HUD) provides critical flight cues for the pilot, making the job of low-level, contour flying a bit easier. The SuperCobra is the only attack helicopter with a dual antiarmor capability, which means it can be equipped to fire either the TOW or the laser-guided HELLFIRE missile. There is a 5-inch cathode-ray tube (CRT) display for control of these missiles. Cockpit instrumentation can be read by pilots wearing night vision goggles.

In addition to a dual antiarmor capability, the USMC is considering a dual air-to-air capability. This would help secure the AH-1W's own protection and contribute to safer escort for troop-carrying helicopters. The SuperCobra is already configured to carry two AIM-9L SIDEWINDER missiles, making it the first production helicopter in the Free World armed to fight enemy aircraft. However, for close-in engagement, the Marines would like to have the option of attaching the lighter-weight STINGER missile.

The AH-1W SuperCobra is an extremely effective and versatile attack helicopter, and plans to incorporate night targeting systems, including laser designator, high resolution Forward Looking Infra Red (FLIR) and Doppler navigation, promise to make the Cobra family tree branch out even further.

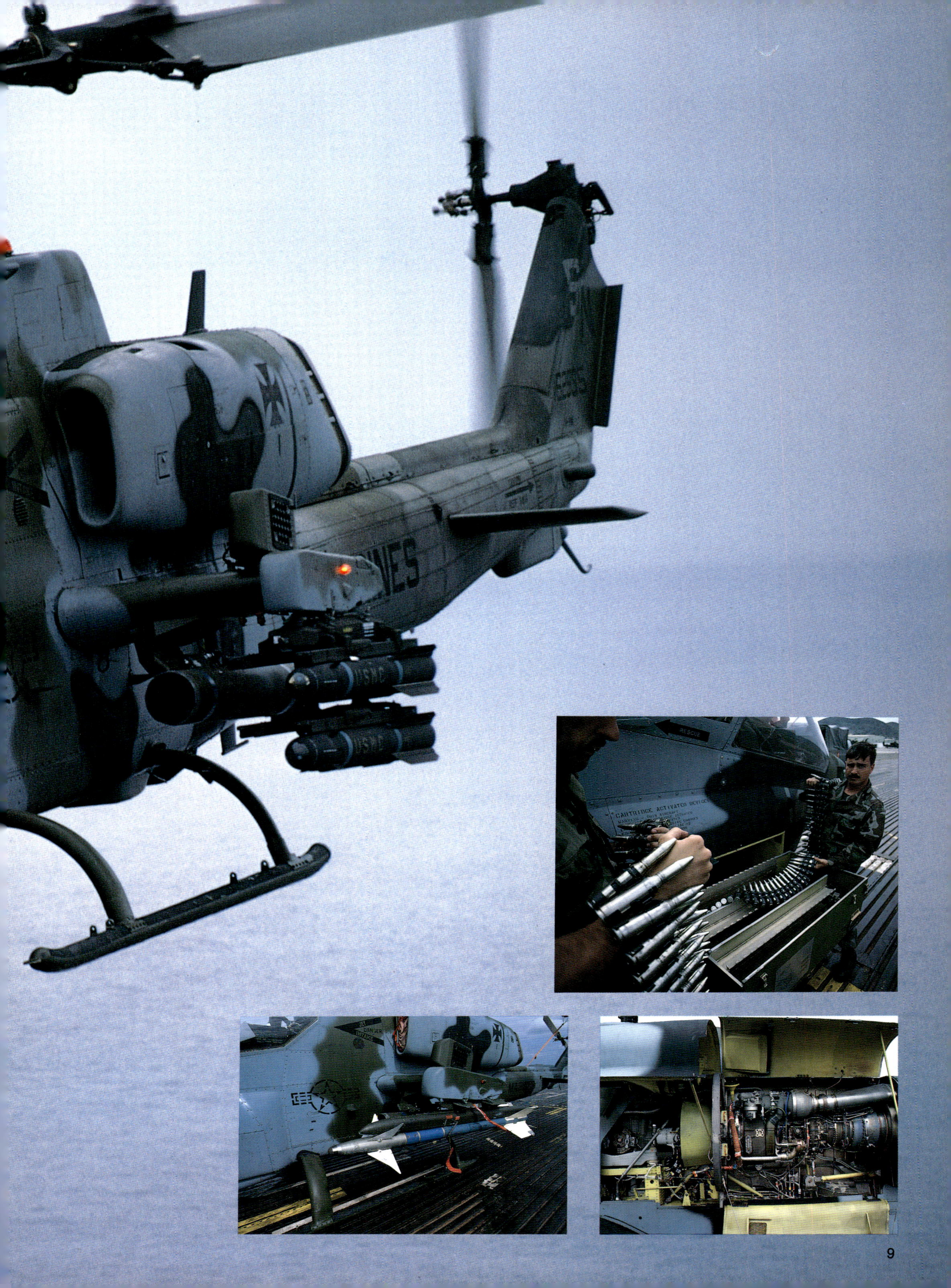

ATA: HELICOPTER

"The presence of opposing helicopter forces on the battlefield makes ATA (air-to-air) combat inevitable."

US Army Field Manual FM-1-107 (Air-to-Air Combat)

Armed helicopters have been used since the 1950s, but only in the past five years has much attention been paid to the possibilities of helicopters fighting other helicopters. Helicopter dogfighting was previously ignored since so many of the wars in which helicopters played a major role, such as Algeria and Vietnam, were totally one-sided as far as helicopters were concerned. But these days, even many Third World armies are acquiring helicopter gunships. ATA, air-to-air helicopter fighting, is becoming more and more likely.

Wars fought over the past five years suggest the trends in helicopter dogfighting. During the 1982 war in Lebanon, the first known instance of helicopter-vs-helicopter fighting occurred when an Israeli Air Force AH-1S Cobra attack helicopter downed a Syrian Gazelle antitank helicopter by hitting it with a TOW antitank missile. Such encounters were unanticipated, but will grow in frequency. One of the most unlikely examples of aerial combat reportedly occurred during the fighting between Iran and Iraq. According to press accounts, an Iranian F-4 Phantom decided to engage an Iraqi Mi-24V Hind D attack helicopter, only to fall victim to the helicopter's nose-mounted 12.7mm Gatling gun!

The Falkland War saw the extensive use of helicopters by both sides. The Argentinians did have A-109 Hirundo armed helicopters, but these played little role in the fighting. But transport and scout helicopters on both sides were victimized by fixed-wing aircraft. A British Army AH.1 Scout helicopter was shot down by a Pucara ground attack plane, and another Scout was forced to land by Argentinian fighters. A flight of four Argentinian Puma transport helicopters was jumped by a pair of Royal Navy Sea Harriers. One Puma crashed trying to evade the Harriers, and two others landed and were destroyed on the ground by the fighters. The Falkland War provides ample warning that an important element in helicopter ATA combat is the need to take defensive measures against fixed-wing aircraft.

Both the U.S. Army and the U.S. Marine Corps have examined the lessons of recent air combat and have begun to take steps in the areas of training and equipment. The quote at the beginning of this article is taken from the U.S. Army's first field manual to discuss, in detail, the nature of ATA tactics. Until the 1980s, the main threat to helicopters was ground fire. But helicopter crews must now pay attention to threats from above as well as the more familiar threats from below. Helicopters have always been vulnerable to fighter aircraft, but recent additions to the Soviet inventory, such as the slow and robust Sukhoi Su-25 Frogfoot ground attack aircraft, increase the risk.

Especially worrisome to NATO helicopter crews has been the growth of a very potent Soviet Air Force attack helicopter force. The best known element of this force is the Mi-24 Gorbach (Hunchback) attack helicopter, more commonly called by its NATO codename, the Hind. The most common variant, the Mi-24V Hind D, is armed with a turreted 12.7mm Gatling gun under the nose, and can carry an assortment of 57mm rockets and guided antitank missiles, all usable in ATA fighting. The Yugoslavs have already modified the 9M32M Strela 2 (SA-7 Grail) antiaircraft missile for helicopter-to-helicopter fighting, and similar Soviet efforts with newer missiles

OH-58D with ATAS launcher

DOGFIGHTING by Steven J. Zaloga

like the Igla (SA-14 Gremlin) seem certain. The Hind F has appeared, armed with twin 23mm autocannon on the left fuselage side instead of the normal nose machine gun turret. While this weapon may be intended for ground attack, a long-barreled weapon of this type offers significant long-range, standoff advantages over the shorter-barreled machine guns characteristic of U.S. attack helicopters. Soviet helicopter tactics also seem to be reacting to the ATA challenge. The Soviets have apparently been experimenting with a three-helicopter team with two helicopters carrying out the traditional ground attack mission and a third Hind staying above as top cover to keep an eye out for antihelicopter threats of both the ground and air varieties.

The Hind will be followed in the next few years by a considerably improved attack helicopter, the Mi-28 Havoc. The Havoc is being accompanied by a variety of new heliborne weapons, including a new helicopter-launched unguided rocket, and a new generation of antitank missiles. While these weapons are not specifically designed for ATA, their higher speeds and longer ranges make them excellent improvised ATA weapons. It would not be too surprising to see a dedicated helicopter ATA missile appear.

Few Soviet programs have spurred American efforts in ATA more than the forthcoming Soviet Kamov Ka-36 Hokum attack helicopter. Some U.S. analysts believe this new helicopter is specifically devoted to hunting NATO helicopters, especially the much-respected antitank versions. It is entirely possible that Hokum is only a prototype, or may be intended for other roles, such as scouting. Regardless of Soviet plans for this helicopter, it has been one of the main catalysts in U.S. ATA programs.

The technological response has been reasonably quick by U.S. Army standards. The successful STINGER manportable antiaircraft missile has been adapted as the ATAS (air-to-air STINGER) for use as a helicopter self-defense weapon. The STINGER is attractive in this role for a variety of reasons. To begin with, it is relatively light, which means it can be carried on smaller helicopters, including scouts like the OH-58. A lightweight launcher for two STINGERs, and the necessary electronic adaptors and sights weigh only 123 pounds. Typically, a helicopter would carry a single launcher with two missiles. The STINGER has a small warhead compared to most air-to-air missiles, but it has proved to be adequate when used against Soviet Hind helicopters in Afghanistan.

The U.S. Marine Corps has shown an interest in a more potent weapon with a better kill probability against any high-performance jet fighter that should attempt to molest Marine helos. The upengined AH-1W SuperCobra is capable of carrying two SIDEWINDER missiles. These SIDEWINDERs are of the all-aspect variety that proved so successful in Israeli hands in 1982 over Lebanon and in British hands over the Falklands. They allow a helicopter to engage an enemy fighter, even in a direct nose-to-nose engagement.

McDonnell Douglas has shown an AH-64A Apache fitted with SIDEWINDER. But the decision to outfit Army attack helicopters with such a large weapon raises the valid concern of many foot-soldiers over the role of the attack helicopter. Officers in mechanized units have come to depend on attack helicopters as their own pocket air support, available when the Air Force is out hunting MiGs. There is some concern that giving an Apache pilot SIDEWINDER will provide the irresistible urge to go out Hind-hunting, to the neglect of the more mundane and dangerous work of hunting out enemy tanks.

AH-1W equipped with SIDEWINDER

Every SIDEWINDER carried decreases the useful combat load of antitank munitions by over 170 pounds.

The weapons already available to the attack helicopter should not be overlooked. The most potent one is the laser-guided HELLFIRE antiarmor missile. The HELLFIRE can "reach-out-and-touch-someone" at over 5,000 meters (3 miles). Not only does it have better range than the TOW, but it is much faster, and has a much larger warhead. No amount of helicopter fuselage armor will stop the HELLFIRE's 20 pounds of high explosive. The HELLFIRE is not a fire-and-forget infrared guided missile like the ATAS. The Apache crew must keep the laser sight on the enemy helicopter until impact. This disadvantage is outweighed by its much larger warhead, which is over 10 times larger than the STINGER's.

At close ranges, existing helicopter autocannon are lethal, even against the armor on the Mi-24 Hind. The Apaches' 30mm autocannon and the 20mm cannon on late-model AH-1 Cobras are very lethal weapons in ATA engagements. Helicopters have an added advantage in dogfighting in that their weapons are mounted in traversable barbettes that can be aimed by the gunner while the pilot flies the helicopter. The 2.75-inch FFAR/Hydra 70 unguided rockets can also be used in ATA combat. While not as predictable as guided antitank missiles, their 70mm high explosive warheads would make fast work of a rotor blade or cockpit. A salvo of rockets would compensate for the inaccuracy of a single rocket by sheer weight of numbers.

Concerns over future ATA engagements are affecting U.S. Army helicopter plans. The Army's new LHX (light helicopter experimental) program is one of the clearer examples of this. One version of the planned LHX, the SCAT, would be a high-speed, armed scout/antitank helicopter. It is clear from the requirements of this helicopter that ATA fighting is envisioned as one of its major roles. The main improvements sought over existing helicopters are greater speed and maneuverability. Soviet attack helicopters have traditionally been faster than comparable U.S. helicopters, and there is special concern in this area due to the possible arrival of the Soviet Hokum helicopter on the scene. LHX would probably be the first NATO helicopter developed from the ground up with the ATA role in mind.

Another important example of the influence of ATA can be found in U.S. Army plans for an upgraded AH-64 Apache, called the AH-64B. The new version of the Apache would incorporate a number of changes in the fire control and weapons systems, optimized for ATA fighting. The Army has not yet made a commitment to this program, but McDonnell Douglas may fly a prototype by the end of the decade. One of the main challenges will be to design a system suitable for locating and tracking opposing helicopters which can be managed by a two-man crew along with all their other responsibilities. Helicopter crews do not only have to worry about their target; in the low altitude environment, they have to pay attention to the threat posed by enemy air defense missiles and guns below them. Significant improvements will have to be made in cockpit design to permit a crew to keep track of all the threat warning and target acquisition data simultaneously.

Programs like the AH-64B Apache are also based on the premise that ATA may be a better approach to air defense than the traditional one of relying solely on ground-based guns and missiles. Due to the use of long-range standoff guided antitank missiles, it is very hard for ground-based gun and missile systems to engage attack helicopters. Instead of waiting in vain for the enemy attack helicopters to come into the range of ground-based air defense, advocates of this alternative suggest that ATA dogfighting helicopters could go out and get enemy antitank helicopters where they lurk. An antitank helicopter hovering in a treeline five kilometers back from the battleline is a nearly impossible target for radar-based guns and missiles. But it does not pose such problems for an attack helicopter. It remains to be seen whether the Army will broaden its thinking about the means of obtaining air defense against attack helicopters, and whether technological improvements will make helicopters a viable partner in the air defense of ground units.

There's an old adage in the aerospace business that "when you don't know where you're going, any route is just fine." At the moment, ATA combat is in this predicament. There is very little certainty whether ATA helicopter combat will be a major factor in the combat employment of military helicopters. There has never been a major conflict involving large numbers of strike fighters and attack helicopters to act as a guide. Recent wars provide only a very shaky forecast of the likelihood of helicopter dogfighting. But it is prudent to contemplate such possibilities. To their credit, both the U.S. Army and U.S. Marine Corps have begun to do so.

Sikorsky CH-53D/E

The CH-53E Super Stallion, built by the Sikorsky Aircraft Division of United Technologies, is the latest version of the U.S. Marine's heavy assault-support and troop transport helicopter. Entering service in 1981, it replaced the earlier CH-53A/D Sea Stallions.

The new Super Stallion adds a third (uprated) General Electric (GE) T64-GE-416 turboshaft engine, each producing 4,380 shaft horsepower (shp). Improvements for the E-variant include the addition of a seventh titanium-spar blade to the 79-foot diameter main rotor array. The transmission has also been uprated to match the 13,140 shp of the engines. The helicopter's 20-foot diameter tail rotor has a new configuration to enhance control. These modifications give the 196 mph Super Stallion the capability of lifting 16 tons (compared to five tons for the earlier model).

The CH-53E is used to transport troops, equipment and supplies for a

landing force during ship-to-shore movement and within the objective area. With its uprated heavy-lift performance, the CH-53E has a 32,000-lb cargo capacity, hauling fully 93 percent of a Marine ground division's equipment (compared to 38 percent for the CH-47A/D's). It can also lift or recover virtually all Navy and Marine Corps fighter, attack and electronic warfare aircraft.

A rear-loading ramp and cargo winches, roller conveyors and tie-down fittings ensure fast, safe, internal handling of transported vehicles and supplies The 33,826-lb helicopter can haul 56 troops or 24 litter patients. For self- protection, the Super Stallion can be armed with twin 7.62mm M-60 machine guns or two .50 caliber M-2x.

Fully loaded, the CH-53E has a combat radius of 50 nautical miles. There are twin 650-gallon fuel tanks on both main landing gear sponsons and a probe which allows for in- flight refueling.

15

APACHE
AH-64

17

APACHE — ARMOR KILLER

by Joe Poyer

In 1881, the U.S. 6th Cavalry fought the Apache Indians, led by Geronimo, Chato, Nana and Juh, throughout Arizona. Exactly 105 years later, the 3rd Squadron, 6th Cavalry Brigade became the first combat ready unit to field the AH-64 helicopter built in Mesa, Arizona and named in honor of the Apache warrior. The AH-64 Apache is the first and, so far, only day/night, adverse weather, antiarmor battle helicopter in the U.S. Armed Forces.

The Apache was designed and built by Hughes Helicopters, Inc. which was acquired by McDonnell Douglas Corporation in January 1984 and became the McDonnell Douglas Helicopter Company in August 1984. The Apache was designed to meet the threat of advanced Soviet main battle tanks (MBTs). It is also being given air-to-air combat capability to deal with the threat posed by a new generation of Soviet helicopters — the Mi-28 Havoc and Hokum which will provide air-to-air protection for Soviet armored formations.

The Apache is well equipped for either of these tasks. In the antiarmor role, it carries up to 16 laser-guided HELLFIRE antitank missiles which have a range beyond three miles and can penetrate all known MBT armor.

In the air-to-air role, Apache can reach out and swat its attackers with the McDonnell Douglas turret-mounted M230 30mm Chain Gun, which is controlled by the crew's "Luke Skywalkerish" helmets. Wherever the pilot looks, the muzzle of the M230 follows. Apache can also provide suppressive fire against light armored vehicles and troop concentrations. For this task, it is fitted with as many as 76 2.75-inch (70mm) folding fin aerial rockets that can deliver antipersonnel mines or high explosives, backed up by its 30mm gun. Two men can refuel and rearm, the helicopter's weapons systems within 10 minutes maximum.

The Apache grew out of the concern following the Vietnam conflict that the U.S. Army needed a helicopter specifically designed for the attack role. Soviet armored capability in Europe was growing and previous helicopter "gunships" were really only flying platforms on which various weapons had been mounted. The Army issued a proposal request for an advanced attack helicopter designed from the ground up to serve as an attack vehicle — one in which the weapons systems were an integral part of the design and not add-ons.

Following an extensive evaluation of several designs and a flyoff with a Bell Helicopter design, Hughes Helicopter's YAH-64 prototype, based on their Model 77, was chosen. Hughes earned a development contract in December 1976. The first production Apache was rolled out on September 30, 1983 and was delivered to the Army the following year, on January 26. The Army has contracted to buy 593 Apaches with deliveries scheduled through June 1990. A total of 286 Apaches have been delivered, as of the end of October 1987, and McDonnell Douglas is building more at the Mesa facility at the rate of 10 per month, each with a "fly away" cost of $9.9 million.

The Apache was designed to fly to a combat area at altitudes measured in tens, rather than thousands, of feet using a technique called nap-of-the-earth (NOE) flying. In this way, the aircraft will be able to stay well below the level of most antiaircraft defenses of the Soviet OMG. Apache, like its namesake, is adept at using every bit of natural cover to ambush its foe - enemy armor.

The AH-64A Apache's primary mission is antitank warfare. The Soviets have at least 17,000 T-64, T-72 and T-80 MBTs out of a total of some 52,000 that present a major threat to NATO forces in the event of a war in Europe. The T-72 and T-80 tanks are equipped with advanced armor of the latest design whose thickness exceeds 650mm and, probably, 700mm. Soviet tactical doctrine calls for the use of tanks with fast moving, long range strike forces called Operational Maneuver Groups (OMGs). These OMGs will field 600 or more tanks to engage targets 300-400 miles deep in NATO territory. They will be protected by the world's heaviest concentration of air defenses.

Its onboard "eyes," the Target Acquisition and Designation Sight/Pilot Night Vision Sensor (TADS/PNVS) system, manufactured by Martin Marietta, feeds the pilot's and copilot/gunner's (CPG's) integrated

helmet display and sighting system (IHADSS). Built by Honeywell Avionics Division, the IHADSS enables Apache to identify its targets and establish an order of priority. For example, its priority might focus on an antiaircraft gun or missile defenses first, then MBTs one, two and three next. The crew will then define the target's geographic coordinates relative to the Apache's position and light it up with its laser rangefinder/designator.

The decision to fire one or more HELLFIRE missiles one at a time, in salvos or in ripples can be made by the pilot or CPG with assistance from the Fire Control Computer (FCC). While the missiles are enroute, they search for reflected laser light. If the area is too hot for Apache to loiter while the missiles are in flight, control can be handed off to a second aircraft, or even to a ground unit equipped with a compatible laser sight/ designator.

Order of priority does not necessarily mean one missile at a time. Apache can handle up to 16 different HELLFIRE targets in a single mission — "lock on after launch" capability. An antiaircraft gun or missile launcher might be designated as the primary target but subsequent missiles, fired singly, in salvos or ripples, will home in on the other targets seconds later.

Recently, the Army and Marine Corps demonstrated another unique ability when 13 AH-64A Apaches from the 1st Squadron, 6th Cavalry, Fort Hood, Texas and two AV-8B Harrier IIs from Marine Attack Squadron VMA 542 at Cherry Point, North Carolina worked together to destroy ground targets. An Apache hovered at treetop level and designated targets with its laser while firing 2.75-inch rockets to suppress ground defenses. A Harrier swept in low, popped up, acquired the laser target in a handoff, released its bombs and banked away sharply to avoid enemy fire. During one mission, six attack runs were made without a miss using "dumb," or unguided, bombs. The same tactics have also been worked out with Air Force A-10 ground attack aircraft and F-16s at Nellis Air Force Base (AFB), Nevada, and with Navy A-6s at Fort Lewis, Washington.

ILLUSTRATION BY JOHN BATCHELOR

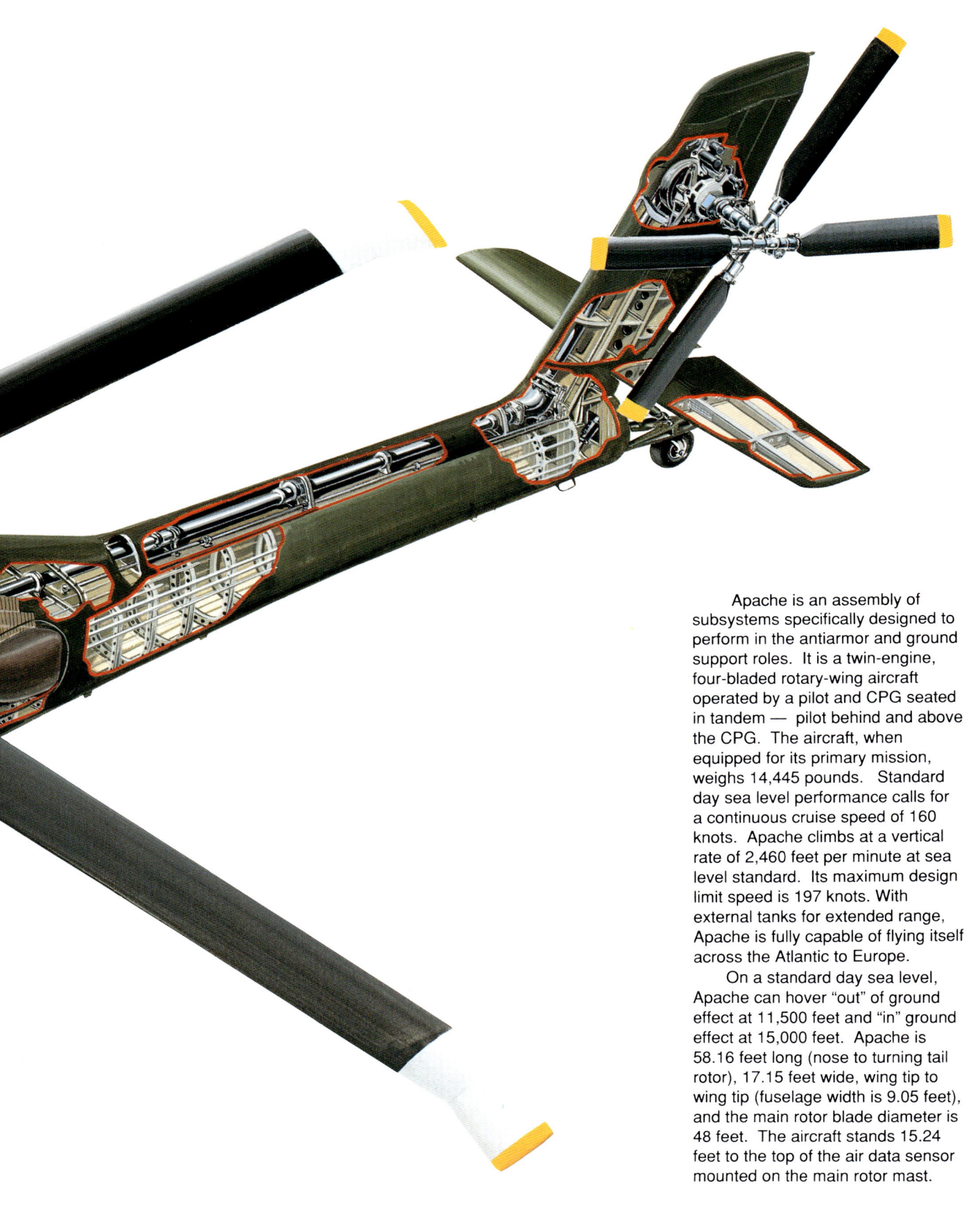

Apache is an assembly of subsystems specifically designed to perform in the antiarmor and ground support roles. It is a twin-engine, four-bladed rotary-wing aircraft operated by a pilot and CPG seated in tandem — pilot behind and above the CPG. The aircraft, when equipped for its primary mission, weighs 14,445 pounds. Standard day sea level performance calls for a continuous cruise speed of 160 knots. Apache climbs at a vertical rate of 2,460 feet per minute at sea level standard. Its maximum design limit speed is 197 knots. With external tanks for extended range, Apache is fully capable of flying itself across the Atlantic to Europe.

On a standard day sea level, Apache can hover "out" of ground effect at 11,500 feet and "in" ground effect at 15,000 feet. Apache is 58.16 feet long (nose to turning tail rotor), 17.15 feet wide, wing tip to wing tip (fuselage width is 9.05 feet), and the main rotor blade diameter is 48 feet. The aircraft stands 15.24 feet to the top of the air data sensor mounted on the main rotor mast.

AIRFRAME — The aircraft's structural design emphasizes crash and enemy fire survivability. To achieve as much protection for the crew as possible, the fixed landing gear absorbs a straight down impact at up to 20 feet per second, the airframe collapses and the crew seats provide additional protection. The crew has a 95 percent probability of walking away from a 42 feet per second vertical crash impact. While in NOE flight near Fort Hood, a 6th Cavalry Apache struck power lines during a tactical mission. Not only did the crew walk away from the crash that could have killed them in any other helicopter, but the Apache was repaired and flying again within six months.

AERODYNAMIC SURFACES — The Apache has two stub wings that serve as attaching points for four external pylons to carry HELLFIRE or 2.75-inch rockets, and for external fuel tanks. Tail surfaces are a fixed vertical and an articulated horizontal stabilizer — dubbed a stabilator — which is mounted aft of the vertical stabilizer and moves as one piece.

FUEL SYSTEM — Fuel cells are located fore (156 gallons) and aft (220 gallons) of the ammunition bay and low in the fuselage. They are self- sealing against 12.7mm rounds and can absorb the impact of rounds up to 23mm. Against 14.5mm armor piercing rounds, they will self-seal to provide a 30-minute fuel supply. Nitrogen gas purging prevents fire in the event of incendiary round penetration. External tanks can be mounted on the wing pylons for extended range.

ROTORS — The Apache main rotor is a fully articulated four- bladed assembly. Each blade is built around a four-cell structural box of stainless steel spars and fiberglass tubes. A heavy gauge stainless steel leading edge can withstand tree branch strikes up to two inches in diameter. The blade structure can withstand damage from 12.7mm machine guns or 23mm high explosive (HE) shells. The tail rotor is composed of two sets of twin-bladed teetering hubs mounted in titanium forks at 55 degrees from the vertical, which reduces the audible sound signature.

POWERPLANTS — Two General Electric (GE) T700-701 engines — similar to those flying in the UH-60 Black Hawk helicopter — rated for twin-engine maximum continuous power of 1,694 shaft horsepower (shp) each, drive the rotor system through individual engine gear boxes mounted in the nose and a main transmission. A 125-shp auxiliary power unit operates the transmission accessory gear box and pressurized air system to start the engines or to provide full electrical power. The T700-GE-701 is an engine with proven performance. The engines are mounted 6.6 feet apart and protected by armor to minimize enemy fire damage to both at one time.

DRIVE SYSTEM — This includes the engine nose gear boxes, the main transmission, intermediate and tail rotor gear boxes, drive shaft and static masts. Nose, intermediate, and tail gear boxes are grease lubricated and can operate for at least 30 minutes if the lubrication system fails. In tests, the main transmission has continued to run for up to one hour, and the intermediate tail rotor gear boxes up to two and one half hours, without lubrication after being damaged by gun fire.

ARMOR/SURVIVAL SYSTEMS — Apache armor, boron carbide bonded to Kevlar, is distributed throughout the airframe to provide the maximum amount of protection for the crew and vital systems. Transparent and opaque blast shields separate the pilot and CPG and protect against gunfire up to 23mm high explosive incendiary (HEI) so that both crew members won't be knocked out by a single round. Armored seats and airframe armor are rated to withstand rounds up to 12.7mm armor piercing incendiary (API). Those aircraft systems and components not shielded are redundant and separated. Their sensitive components are isolated, and fire suppression systems are employed on the engine and Auxilliary Power Unit. Spalling resistant materials have been widely used in Apache and non-critical items have been positioned to mask those that are. A wire strike protection system will be incorporated. Passive cutters sever power or other lines during low level flights or landings. Wire strikes are a major cause of helicopter crashes.

NAVIGATION SYSTEMS — To assure that Apache will arrive on station at the right moment, a Heading Attitude Reference System (HARS) and an AN/ASN-128 Doppler Radar operate in conjunction with the FCC and an Air Data system to provide precise navigation. The information is fed to the crew on a continuous basis through the TADS/PNVS and IHADSS displays. Check points can be established along the flight route and the helicopter's exact position refined in flight as each check point comes within range of the TADS. This ensures precise positioning and waypoint targeting that is invaluable in bad weather or at night. The PNVS and Forward Looking Infra Red (FLIR) provide the Apache night time NOE capability in the form of high resolution video on the IHADSS and TADS displays.

27

WEAPONS SYSTEMS

Send a tank to kill a tank used to be the rule of thumb. Now you send an Apache. The AH-64A is a veritable flying tank itself. It can carry four HELLFIRE launchers holding a total of 16 missiles or 76 2.75-inch (70mm) folding fin rockets in four launchers.

HELLFIRE — This antiarmor missile is the Apache's primary weapon. Its flight path allows it to attack from above to penetrate a tank's softer turret top. Its single-shot killing power, ability to engage multiple targets and growth potential in terms of range and heavier warheads, are unmatched in the world today. The HELLFIRE can conceivably allow a single Apache to engage an entire tank column using either the indirect fire or ripple fire methods if a sufficient number of laser designators can be deployed cooperatively on the ground or in the air.

M230 30MM CHAIN GUN — This can be operated as the primary or backup weapon and, if the FCC is damaged, the gun can be fired in the stowed position (straight ahead) by aiming the helicopter.

Normally, the gun is slaved to the TADS line-of-sight, constantly updated by the FCC, and fired by the CPG. But in backup, either crew member can fire the Chain Gun when wearing the IHADSS helmet. The crew member has only to look at the target and the gun tracks to his line-of-sight. (This is much better than the old "grease pencil mark on the canopy" method.)

2.75-INCH (70MM) FFAR — The 2.75-inch folding fin aerial rocket (FFAR) can be equipped with a variety of warheads and can be selected and fired by the pilot using his IHADSS; pilot selected and fired; or fired by the CPG using the IHADSS and FCC to provide accurate targeting information. Fuse settings can be made before launch to compensate for range or target heights and the rockets can be launched as singles, in pairs, multiple pairs, or in salvos.

Two battalions of the 6th Cavalry deployed to West Germany during Reforger '87.

As of mid-October 1987, five Army units are now operational with Apache — the 1st, 2nd and 3rd Squadrons, 6th Cavalry Brigade; 5th Squadron, 17th Cavalry Brigade; and the 1st Battalion, 227th Aviation, 1st Cavalry. A 30-man cadre of the 30th Aviation Battalion, North Carolina National Guard — the first Guard unit to receive the AH-64A — is training with the 1/227. The unit has returned to North Carolina to train other Apache crews and ground support personnel. The entire battalion will then return to Fort Hood for final training and evaluation during 1988. Two battalions of the 6th Cavalry deployed to West Germany during the Reforger '87 Exercises. One has remained to form the nucleus of a permanent AH-64A Apache force in NATO.

APACHE'S FUTURE

An Advanced Apache may be in the works. The new version would have updated fire controls for 30mm and a new weapon for increased air-to-air capability — probably STINGERs — improved crew visibility and a new air-to-air missile system. Aircraft survivability would be enhanced through increased redundancy. Advanced flight

controls would improve precision flying and flight path control.

General Electric is developing an updated version of the twin T700-701 engine that could be available within a year. Among other changes will be greater horsepower — up from 1,710 to 2,000 — and an improved electronic control unit to further improve Apache's air-to-air combat capability and provide greater dash speed.

McDonnell Douglas has proposed a sea-going version of the Apache to the U.S. Navy that would operate from cruisers, destroyers and frigates, merchant ships or other auxiliary vessels without aircraft carrier support. Flying in the outer air battle regions, beyond range of the host ship's sensors, it would identify and counter enemy air and surface craft threats.

One of the safer predictions that can be made concerning military weapons systems is that the AH-64A Apache, a system that works, will be flying in various guises and configurations well into the next decade, if not the next century. Geronimo, Chato, Nana and Juh would have been satisfied that their people's name is being carried honorably.

CHAIN GUN®

McDonnell Douglas M230 30mm Automatic Cannon

by Robert Bruce

The world's most advanced attack helicopter is the U.S. Army's AH-64 Apache, armed with a dazzling array of sophisticated and devastating weapons. But, arguably, the most versatile and cost effective of these is its 30mm automatic cannon produced by McDonnell Douglas Helicopter Company. Capable of punching out over 10 rounds per second of specialized ammunition, the lightweight M230 Chain Gun gives the Apache the ability to defeat the widest range of targets on the high intensity battlefield. Unclassified performance figures show that its reach extends to 4,000 meters and beyond, providing a comfortable standoff distance from most front line threats. Its total system weight is a modest 1,509 pounds — even when fully laden with a remarkable 1,200 rounds of ammunition.

A single jam, however, could instantly reduce any such system to an inert 3/4-ton weight penalty. Absolute reliability is mandatory — under all conditions, in any extreme of weather, from searing deserts to the brittle cold of the Arctic. High tech and high reliability must go together in the real world of the modern battle arena.

The Army had begun extensive automatic cannon testing back in the early 1970s with the requirements of its conceptual "Advanced Attack Helicopter" in mind. No effort or reasonable expense was spared to identify the best system for the job. Candidates included conventional self-powered designs, gatling-type revolvers, and an externally driven system of unique configuration from Hughes Helicopters — later to become a key part of McDonnell Douglas.

Typically, large caliber automatic cannon suffer from any number of problems that must be addressed by tradeoffs in design. Self-powered guns utilize the energy given off by detonation of the ammunition to operate all functions including feed, locking, extraction, and ejection. This means their parts must be relatively heavy and resistant to extreme stress. Also, even small variations in the ammunition can cause stoppages.

Gatling-type weapons utilize external power, usually electric motors or hydraulic systems, to rotate a cluster of barrels into firing position. While this promotes both greater reliability and high rates of fire, the penalties tend to be increased complexity, system weight and bulk, plus greater cost.

In contrast, Hughes' engineers offered their single barrel, rotating bolt, electrically driven Chain Gun, so called because of its motorcycle type sprocket and chain operating mechanism. For helicopter mounting, this well thought out combination overcomes virtually every objectionable aspect of competitive systems. The Chain Gun quickly took the lead in testing, demonstrating clear superiority over its rivals when balancing the factors of performance, reliability, system weight and cost.

Use of a single barrel means lighter weight, few moving parts, and consistent accuracy. Unlike revolving cannon, the Chain Gun's fixed barrel is not subject to boresight

AH-64A 30 mm Area Weapon System

System Weight (Total)	1509 lb
System Weight (Empty)	585 lb
Ammunition Capacity (M789)	1200 rnds
Rate of Fire	625 25 spm
Ammunition Handling System	Linear Linkless

30 MM M230A1 Area Weapon Characteristics

Caliber	30mm Ammunition
Combat	ADEN/DEFA/M789/M799
Practice	ADEN/DEFA/M788
Weight	
Receiver (includes motor)	63 lb
Barrel	35 lb
Recoil Adapter	12 lb
Linked or Linkless Transfer Unit	13 lb
Total Gun System Weight	123 lb
Dimensions	
Length	64.5 in
Width	10.0 in
Height	11.5 in
Frontal Area	55 sq in
Barrel Life	to 10,000 rnds
Rate of Fire	625 25 spm
Time to Rate	0.2 sec
Time to Stop	0.1 sec
Clearing Method	Open Bolt
Effective Impulse (with Muzzle brake)	45 lb sec
Dispersion	<2 mil
Power Required	6.5 hp Reliability

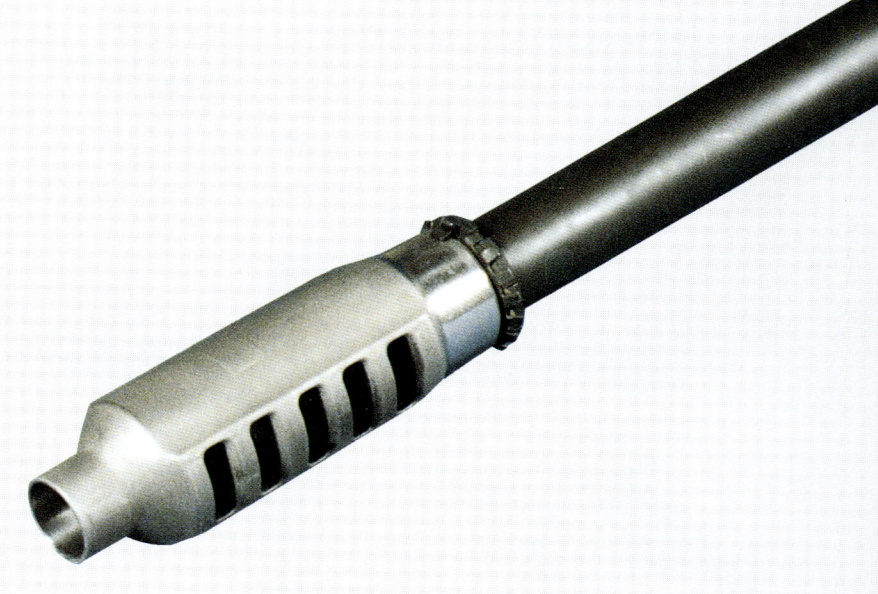

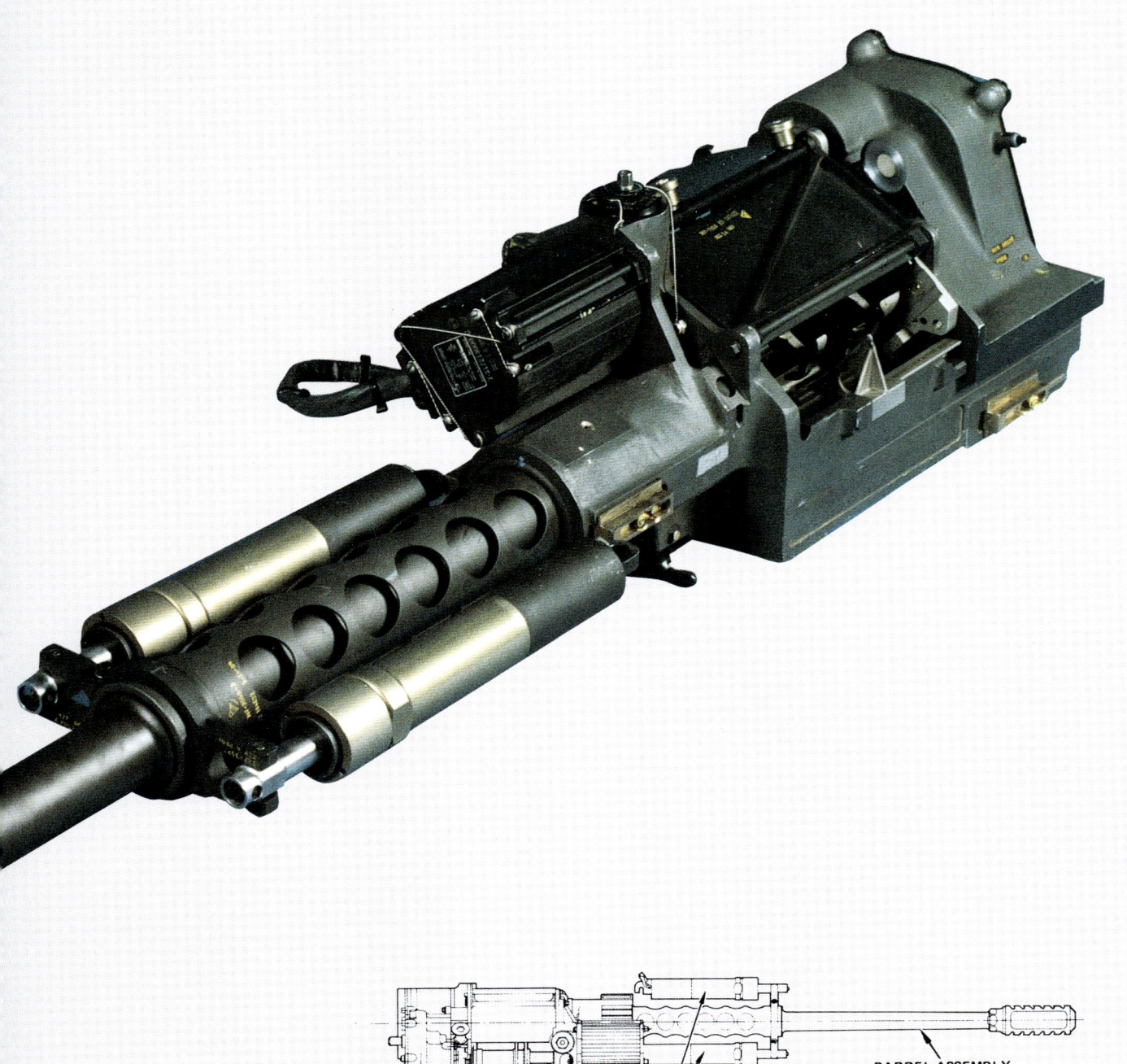

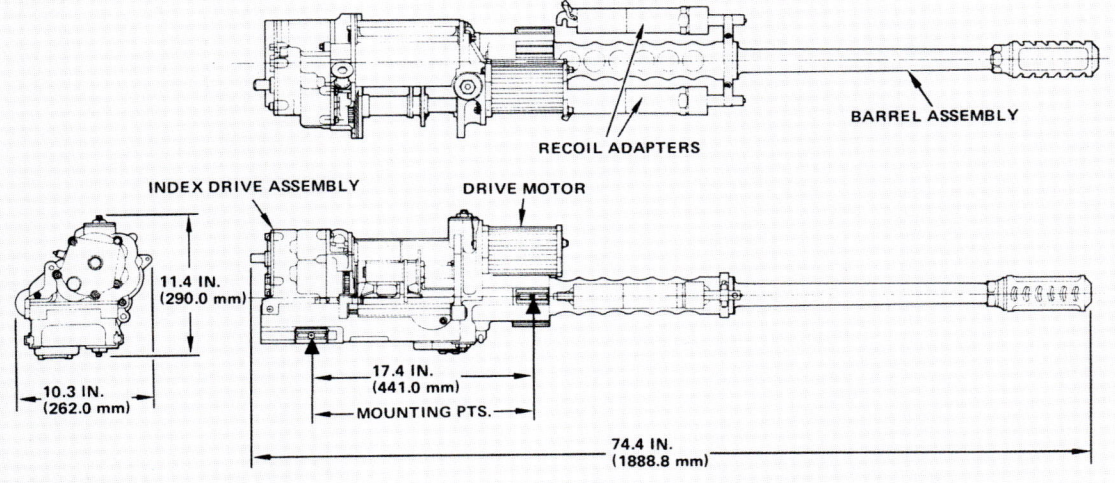

BARREL ASSEMBLY

RECOIL ADAPTERS

INDEX DRIVE ASSEMBLY

DRIVE MOTOR

11.4 IN.
(290.0 mm)

10.3 IN.
(262.0 mm)

17.4 IN.
(441.0 mm)

MOUNTING PTS.

74.4 IN.
(1888.8 mm)

deviations inherent in multi-barrel designs. Use of the attack helicopter's electrical power source is a logical alternative to the problems associated with self-powered systems. Operated by a small 6.5-horsepower motor, the Chain Gun cycles smoothly and precisely with little firing-induced vibration. Damping of the short recoil stroke of the barrel is efficiently accomplished, thanks to twin recoil adaptors and a well designed muzzle brake.

Its deceptively simple mechanism is all the more remarkable for its extraordinary reliability. McDonnell Douglas cites a figure of 100,000 mean rounds between failure. This would be an astounding claim and cause for skepticism were it not firmly backed by contractor and Army testing programs.

Safety in operation need not be compromised for lethality on target. The Chain Gun incorporates features to ensure absolute safety to the air crew and maintenance personnel. When using a simple hand crank for loading and downloading, accidental firing of the gun is blocked by a fail-safe system. Percussion priming of the ammunition means that it is insensitive to detonation from static or stray voltage.

In operation, the gun begins and ends its firing cycle with the bolt fully rearward and no round in the chamber. This eliminates the possibility of accidental explosion of a round left in an overheated chamber. Similarly, "double feed" (attempt to feed a round when there is another in the chamber) is impossible due to a sensing rotor that instantly stops the feed system if a round fails to extract. So, at no point in the feed, extraction, or ejection cycle does the projectile's sensitive fuse come in contact with the mechanism.

AMMUNITION AND FEED

The 30mm M230 series Chain Guns will fire all NATO standard ammunition from Honeywell, ADEN, and DEFA. This gives the system a wide range of options including High Explosive Incendiary (HEI), High Explosive Dual Purpose (HEDP) and inexpensive Target Practice (TP). The HEDP is particularly efficient in that it combines long range with both armor penetration and blast/fragmentation effect. Uploaded with a full belly of 1,200 HEDP rounds, the Apache can easily take on armored fighting vehicles like the Soviet BMP, as well as crew served weapon sites, emplaced troops, or troops in the open. While the Apache is equipped with a light and compact linkless feed system supplied by McDonnell Douglas, the gun can also be fed by linked belts when installed in other airframes or vehicles.

IDEAL FOR HELICOPTERS

Use of advanced metallurgical techniques and rotorwash cooling virtually eliminates overheating of its single barrel even at high rates of sustained fire. The low weight and streamlined design of the gun means a low wind drag factor, ensuring maximum flight performance. All bearings and gearboxes are sealed against grit and moisture, and the system has been proven in testing and in operation to be largely unaffected by temperature extremes. Also, rate of fire control enables the gunner to select the density of each burst to fit the target for conservation of ammunition during each sortie.

ARMORED FIGHTING VEHICLES

The demonstrated superiority of the McDonnell Douglas M230 Chain Gun in the air has carried over to vehicle installations as well. The Bushmaster, a 25mm version, is the standard main gun on the Army's Bradley Fighting Vehicle, the USMC's Piranha, and a number of other Armored Fighting Vehicles (AFVs) worldwide. Sharing most of its predecessor's characteristics, the M242 features a dual feed system that allows the gunner to instantly choose between two types of ammunition for maximum effectiveness against armored or "soft" targets.

Given the obvious merits of the design, the next logical step has been to scale down the Chain Gun to 7.62mm for installation under armor and in aircraft as well. Once again, the efficiency and reliability of the Hughes/McDonnell Douglas chain driven system has been realized in the EX-34 7.62mm Machine Gun.

At least three recurring problems in enclosed guns have been neatly overcome by the smallest and newest member of the Chain Gun family. With relatively long "dwell" (the interval between firing the cartridge and its subsequent extraction) in its firing cycle, the hazardous and uncomfortable buildup of powder fumes in the crew compartment of the fighting vehicle has been virtually eliminated. Also, empty cases are ejected forward and outside the turret so that shell catcher bags may be eliminated entirely. Finally, the EX-34 features a quick-change barrel that allows a swap in less than 10 seconds without the bother of unloading the gun. Hard evidence of the virtues of the EX-34 is at hand in a license agreement with Royal Small Arms Factory — Enfield, England for the manufacture of the gun to be installed in the British MCV-80 armored personnel carrier, and other ground, sea, and air platforms.

With over 15 years of design refinement and field experience, the Chain Gun family of automatic weapons represents the highest standards of quality and effectiveness. Chosen by the world's most technologically advanced country to arm its latest generation of attack helicopters and armored fighting vehicles, McDonnell Douglas continues to demonstrate a leading role in the Free World's defense systems.

Bell Helicopter
OH-58D AHIP Scout Helicopter
by Damian Housman

The armed scout helicopter of today bears little resemblance to the balloon of more than 120 years ago, but they are indeed closely related. Without doubt, modern concepts of aerial reconnaissance owe their existence to the tethered, hydrogen filled balloons of Thaddeus Lowe, who used them to watch and report Confederate troop movements and field activities during the American Civil War.

By today's standards, the observation balloon was inflexible, inefficient, and vulnerable. After all, it was tethered to a single position and could only observe what a man could see from that position. And once the enemy saw the balloon was up, every soldier wanted the honor of shooting it down. Surprisingly, though balloons were extensively used during the Civil War, none were shot down, despite many attempts.

Observation balloons again proved their worth in the Spanish American War, and finally in World War I. By then, the telephone had replaced the telegraph. And, when the airplane joined the balloon in the air, the "scout" mission was born. The airplane could range over the entire battlefield and beyond, permitting commanders to know what was going on well behind enemy lines.

World War II saw further refinement in the scout mission with the use of the L-4 Piper Cub, a two-seat, reciprocating engine plane which is still a common sight in general aviation. It allowed more accurate adjustment of artillery fire at longer ranges. The crew now had maps, binoculars, and voice radio, which meant that artillery could be quickly redirected.

After the war, the Stinson L-5 became the staple of the scout force, augmented in 1948 by the newer, smaller Aeronca L-16. During the Korean War, these two planes did the bulk of the Army reconnaissance, artillery observation, and liaison, though they were rapidly being replaced by the Cessna L-19 Bird Dog. Redesignated the O-1 in 1962, the Bird Dog was the last light, fixed-wing airplane bought by the Army for this mission.

The Army started buying light helicopters in 1948. The Bell H-13 and Hiller H-23 were bought in the first year of the Korean War, to be used mostly for medical evacuation and administrative support. Though these helicopters were able to hover and land virtually anywhere, they were not as reliable or maintainable as the L-19, and were not deemed suitable for tactical missions. It wasn't until the mid-1950s that the concept of using helicopters in a scouting role was developed. The idea may have languished on designers' drawing boards but for the Vietnam War, which made procurement of a light observation helicopter imperative.

The first Vietnam era scout chopper was the OH-6A, the military version of the Hughes Helicopter (now McDonnell Douglas Helicopter) Model 500D, which arrived in 1965. This was followed in 1969 by the Bell Helicopter OH-58A Kiowa, military counterpart to the Bell Model 206 Ranger, which was more rugged, though less responsive. Both helicopters served with distinction in Vietnam and, between 1966 and 1973, 3,613 light helicopters of both types were delivered.

The experience of Vietnam provided the Army with a conceptual proving ground for advanced scout helicopter tactics and techniques. Further, it gave the Army a firm idea of what would be required of the scout on tomorrow's battlefield and better defined its mission.

A new scout helicopter must fly fast at nap-of-the-earth (NOE) altitudes, locate the enemy, make use of terrain for survivability, call in friendly artillery or helicopter gunship support, coordinate the attack, designate the target with a laser so the gunship's missiles will hit the mark, and do it all regardless of light or weather conditions.

Late in 1980, the Army announced its requirement for a new scout helicopter under the Army Helicopter Improvement Program (AHIP). A year later, Bell's OH-58D was chosen the winner of the competition. Forty-nine months later, in late 1985, the first Bell OH-58D AHIP scout was delivered. Four years is a remarkably short development period these days, when well-run programs average about 10 years. It might be imagined that the new product, therefore, is simply an improved version of the A model. Though there is a surface similarity between the two, the OH-58D Advanced Scout is a revolutionary machine. Far more powerful and maneuverable than its earlier cousin, the Army's new scout is one of the most electronically advanced aircraft flown today.

The OH-58D is powered by a 650 shp Detroit Diesel Allison 250-C30R engine. The main transmission is a 635 Peak HP Bell design with "run dry" capability. The main rotor and tail rotor systems are of composite design with elastomeric bearings. There is minimal vibration. The aircraft is designed for precise control during NOE flight, and can maintain heading control in winds up to 35 knots from any direction. Top forward speed is 119 knots, and vertical rate of climb is 500 feet per minute at 4,000 feet altitude and 95 degrees Fahrenheit (F). Endurance is 2.5 hours.

The cockpit is the first fully integrated, multiplexed cockpit in the Army. The entire cockpit -- display function, sensors, and radio selection, can be controlled by switches on the crew's hand grips. Two multifunction displays present the vertical and horizontal situation, including heading, altitude, airspeed, rate of climb, aircraft position in relation to programed course, waypoint locations, and time, distance and course to selected positions. Five different radios are selectable from the pilots collective, with a total of 35 pre-programed frequencies. Any other frequency desired may be entered using the keyboard. A major function of the displays is to present video and symbology from the mast mounted sight (MMS), including television, thermal imaging, and laser range.

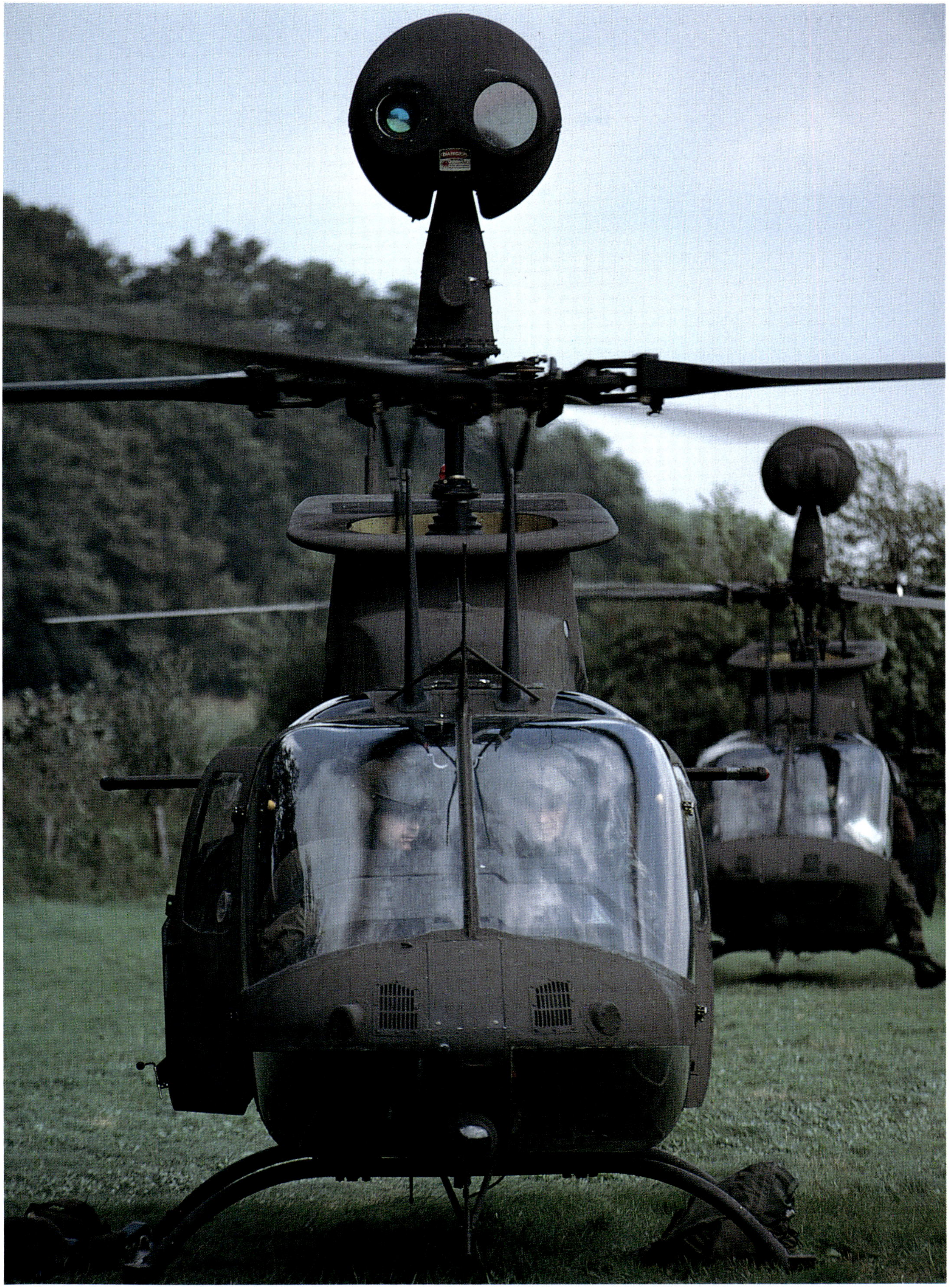
DANGER

The entire cockpit is designed to reduce the workload of the crew in what must be one of the most demanding missions in Army aviation. The cockpit display system is manufactured by Sperry Flight Systems.

The MMS is the most unusual looking aspect of the OH-58D. Looking like a big ball atop the main rotor, the McDonnell Douglas Astronautics Company's MMS is home to the sensor system which makes this the most effective scout helicopter in the world, day or night. Operated by the aerial observer in the left seat, the system searches at at long range for area or point targets.

By day, a television gives the observer a two-by-eight-degree field of view. The system can designate a target for gunship missile attack with either TV or thermal imaging.

Tracking with the MMS can be done automatically, and it is capable of locating and tracking a target from an offset aim point. The laser doubles as a rangefinder.

Locating the MMS on top of the main rotor reduces the vulnerability of the helicopter to enemy fire. The OH-58D needs to expose only the MMS in order to find, track and designate. The helicopter itself can remain safely behind treeline or ground contour. While practically

invisible, the OH-58D can designate for HELLFIRE missiles, COPPERHEAD artillery rounds, and laser-guided bombs, or can handoff the target to an AH-1S Cobra to kill with any of its weapons.

This handoff capability means that the OH-58D can acquire and track a target, and let other helicopters or the field artillery know the precise location. The Airborne Target Hand-over System passes this vital information by secure data burst radio transmission. To give an accurate target position, the OH-58D must have precise knowledge of its own position, which it has through the Attitude Heading Reference

System. This inertial system can be updated by the crew through the MMS, and has a doppler radar backup.

The systems of the OH-58D make it the first Army scout aircraft capable of much more than simply seeing what one man can see through a pair of binoculars. As a member of the attack team, it literally orchestrates the battle by finding and tracking targets and handing them off to AH-1S and AH-64 attack helicopters. The OH-58D can illuminate targets for the Apache's HELLFIRE missile, or for the Cobra's laser tracker to acquire and cue the TOW missile sight to the target. The OH-58D assigns targets, sectors, and kill zones, and allows these antiarmor weapons to be used to their maximum standoff range.

In an air cavalry role, the OH-58D fills many of the same functions as with the attack team, but there is a greater emphasis on reconnaissance and collection of intelligence. The scout will also have an air-to-air capability when the new STINGER missile kit modification is added.

Scouting for field artillery, the OH-58D designates fixed or moving targets for conventional and laser-guided rounds such as COPPERHEAD. Airborne target hand-over data is received, in an instant, directly into the TACFIRE artillery system. Accurate fire the very first round means surprise and devastation for targets which in the past may have been relatively secure.

Brand new, the OH-58D is already making its presence felt in Army aviation units. As of October 1987, 80 were delivered from the first three production lots, out of 99 ordered. A fourth lot of 36 aircraft has been ordered, with a stated Army requirement of 585 units.

Though the OH-58D looks a lot like its predecessor, it represents a giant leap in capability. It is a genuine force multiplier, enhancing the capabilities of other attack, artillery, and observation systems, and provides further evidence of the vast effort underway to modernize the fighting capability of the U.S. Army. ID

SH-60B

The Navy SH-60B Sea Hawk is the air-vehicle portion of the LAMPS Mk III (Light airborne multi-purpose system), for which IBM is the prime contractor. Built by the Sikorsky Aircraft Division of United Technologies Corporation, the SH-60 is a multi-role helicopter used for antisubmarine warfare (ASW), antiship surveillance and targeting (ASST), search and rescue (SAR), casevac, and vertical replenishment at sea. In addition, the SH-60B has the ability to provide early warning against aerial attack. Future versions will be equipped with antiship missiles and torpedoes. This versatile airframe, with its long range and ability to land on small ship platforms, greatly extends the antisubmarine capability of the U.S. surface and submarine naval fleet. The SH-60B is definitely a helicopter on the front line of defense.

43

IBM LAMPS Mk III

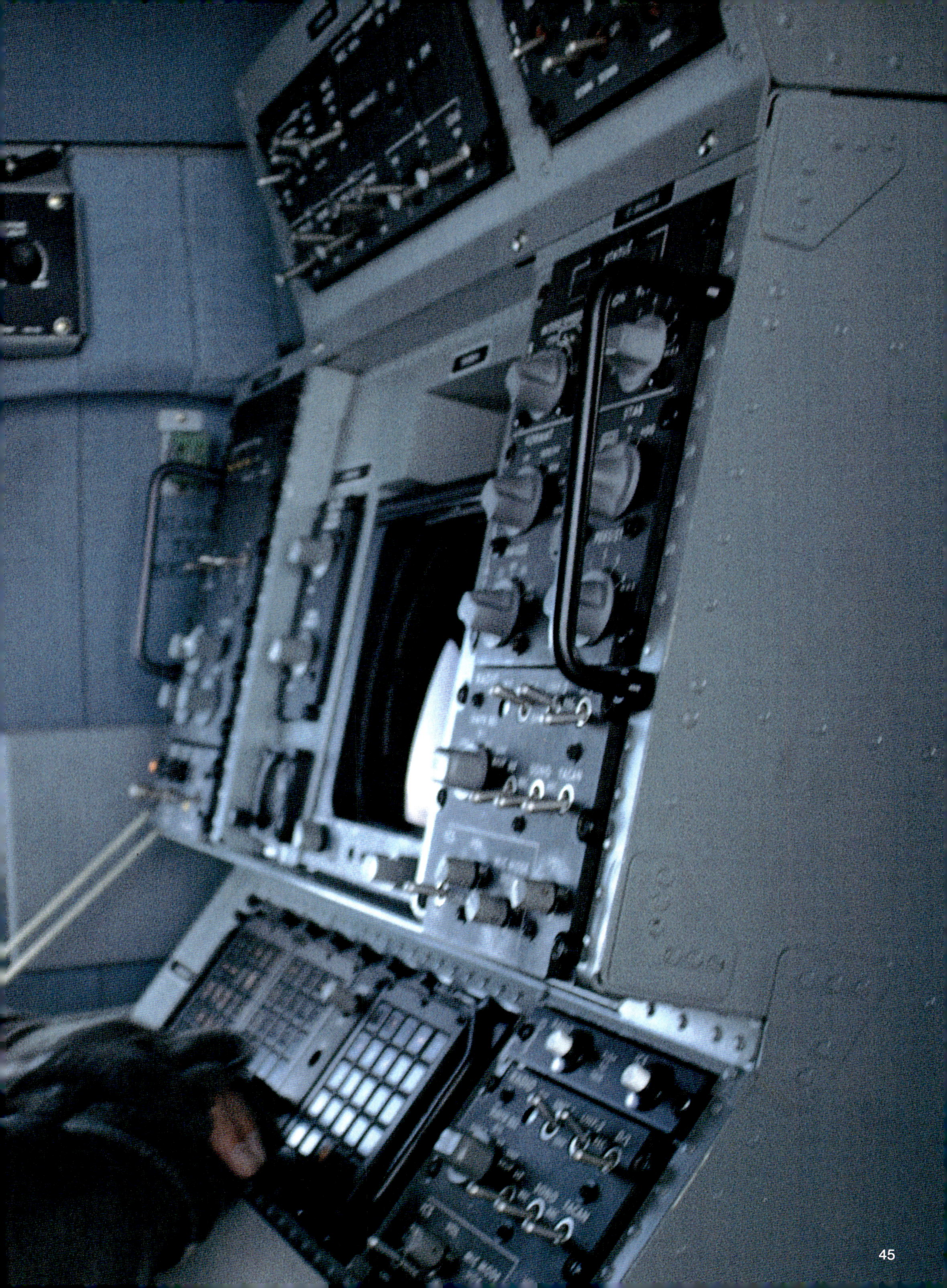

NATO'S COMBAT HELICOPTERS
by David C. Isby

The attack helicopter is becoming an increasingly important element of NATO's force structure. Confederate General Nathan Bedford Forrest's homespun summation of the art of war — "git thar the fustest with the mostest" — underlines both the importance and the challenge of NATO'S helicopters; they certainly have unmatched battlefield mobility, but whether their numbers and tactics will permit this mobility to be decisive is unknown. Helicopters are expensive, and it remains to be seen whether the NATO defense budgets of the late 1980s will permit the promise of these helicopters to be realized.

The increasing Soviet armored threat to Europe in the 1970s and 1980s has led to NATO examining different ways to counter the Soviet superiority in tanks. The mandates of Flexible Response, NATO strategy since the 1960s, means that the nuclear strikes once planned in the late 1950s and early 1960s to halt Soviet thrusts could not be counted upon. The type and strength of conventional forces necessary to counter the Soviet conventional strength on an equal basis, such as the 96 division force envisioned at the 1952 NATO Lisbon conference, were simply not financially sustainable. Faced in the late 1960s and 1970s with having to do more with less in the conventional defense of Europe, NATO looked to the attack helicopter as one of a spectrum of solutions. The helicopter was seen as a way to bring the West's superior weapons technology to bear against Soviet armored strength.

Each nation within NATO has a different approach to the use of attack helicopters. This has led to the development of different types of helicopters as well as different organization and tactics. These reflect not only the overall tactics that each army will use in combat but the strengths and limitations of each type of helicopter. They also reflect the political difficulty - some would say impossibility - of NATO standardizing on major weapons systems. Recognizing this, NATO has tried, even if it cannot standardize equipment, to at least standardize procedures. This way, for example, U.S. Marine Corps helicopters could be assigned to support a hard-pressed Danish Army unit, or Royal Air Force helicopters could be assigned to pick up and move a West German battalion, with all parties knowing what needed to be done.

While the wealth of U.S. combat experience from Vietnam still guides much of NATO's thinking on helicopter operations, the U.S. approach to attack helicopters has not proven to be a model that many of its allies wish to adopt. This is not a reflection of perceived ineffectiveness of the U.S. attack helicopter force, which is widely respected by its NATO allies, but to the cost of the large, multi-purpose AH-1 Cobra and AH-64 Apache attack helicopters. Even the largest of the current European attack helicopters, the A 129 Mangusta, is less than half the weight of an AH-64. Here, capability and cost have had to be traded off against each other. While U.S. attack helicopters must be capable of carrying out a wide variety of missions worldwide, NATO armies optimize their attack helicopters for antitank tactics as part of a defensive strategy. There is also natural pressure to "buy local," so major NATO armies use indigenous designs. The rising cost of research and development, however, may mean that the successor generation to NATO attack helicopters will be produced by international efforts. The Tonal, a joint Italian-British-Spanish-Dutch effort projected for development from the A 129 in the late 1980s, and the new Franco-German efforts, have set the course for the future.

HELICOPTERS AND THE DEFENSE OF THE WEST

A modern army cannot go without attack helicopters any more than it can go without main battle tanks. The helicopter not only provides tactical and operational mobility, but firepower against tanks, air-to-ground guided missiles (ATGMs), machine guns, and other helicopters as well. This is reflected by NATO's increasing use of battlefield helicopters.

The Soviet helicopter threat is growing. New Havoc and Hokum attack helicopters will be serving alongside the Hinds that have been used in Afghanistan. The Hokum is optimized for use against other helicopters with air-to-air missiles. This means that NATO's helicopters will have to evolve an air-to-air role in return or run the risk of conceding much of the advantage that helicopter forces provide. The Soviets have also demonstrated an increased ability for night operations. The quantitative and qualitative improvement in the Soviet helicopter force has broad reaching implications for the conventional balance in Europe. NATO's helicopters will no longer be able to devote their full efforts to offsetting the Soviets' advantage in armor. Rather, they will have to defend themselves against the Soviet helicopter force and try and prevent that force from decimating NATO armor and transport helicopters. Soviet transport helicopters are increasing in numbers and capabilities, with the Mi-26 Halo being the latest example. New helicopters will be required if the Soviets are not to further undercut NATO's technological edge.

To meet the increasing Soviet helicopter threat, a new generation of helicopters is emerging. The U.S. AH-64A is certainly in the lead, but the A 129 will soon join it, followed by the PAH-2/HAX-3/HAP, once its political problems are resolved. The Tonal is unlikely to be in service before the late 1990s.

The advantages in mobility and firepower of the helicopter are tremendous. The late British defense expert, Brigadier Richard Simpkins said "rotor is to track as track is to boot." Helicopters, used in operationally significant numbers, such as those that lift the U.S. 101st Air Assault Division, French 4th Division Aero Mobile, West German Airborne Brigades, and the British 6th Airmobile Brigade, demonstrate an increase in battlefield mobility equivalent to that demonstrated by the Panzer division in 1940.

Attack helicopters, like tanks, must operate as part of a combined armed force. For the attack helicopter, observation and transport helicopters are crucial to effective employment. While the attack helicopter is certainly the cutting edge of the team, it needs to work with the other types of helicopters in combat.

47

How NATO deploys and uses its helicopter forces will help determine how effective its conventional fighting capability will be for the remainder of this century. The late 1970s and 1980s have already seen considerable advances. Even the countries of NATO's Central Front that have been lagging behind in defense spending, such as Denmark and the Netherlands, have recognized the value of the attack helicopter and will be investing some of their scarce procurement budget for numbers of them. But just having the helicopters is not enough. The West must train hard. The relatively few Vietnam and Falklands veterans in helicopter cockpits cannot compare in numbers to the Afghanistan- experienced crews on the other side. Therefore, for NATO helicopters, it is important that frequent and realistic training take the place of combat. This, however, is expensive, requiring money from operations and maintenance funds that are often the first target for budget-cutters. Helicopters, like any other weapon, are only as good as the men that use them. [DI]

WESTLAND LYNX

by Malcolm V. Lowe

The Lynx helicopter, originally produced under an Anglo-French agreement that dates back to the latter 1960s, has grown principally under the design leadership of the British helicopter company, Westland. It is a versatile helicopter which, through the existence of several distinct versions, is capable of satisfying the requirements of both Army aviation and Naval forces. In addition to these proven designs, the Lynx 3 is the more recent dedicated antitank derivative. Other versions, including a tactical transport and logistic support model, have also been developed from the original Lynx design.

The Westland company, an aircraft manufacturer of considerable accomplishment, has now successfully specialized in the helicopter business for more than 30 years. This company has long been the principal supplier of helicopters to the British military. Many of Westland's designs over the years can trace their origin to the United States' Sikorsky company. During 1986, much closer links were forged between Sikorsky and Westland. One result has been the addition of the WS 70 to the Westland group of battlefield helicopters. This aircraft is based closely on the Sikorsky UH-60 Black Hawk. Of Westland's current selection of helicopters, however, the Lynx is very much a European product.

The original prototype first flew on 21 March 1971; the operational Lynx began reaching the British Army and Royal Navy near the end of the 1970s. In its Army (land-based) form, the Lynx serves principally with the British Army Air Corps (with a small number also operated by the Royal Marines). They are easily recognizable by their skid undercarriage, rather than the conventional tricycle-wheeled undercarriage which is fitted to Navy Lynx helicopters. The original production model of the Army Lynx (known to the British forces as the AH Mk 1) was powered by two Rolls Royce Gem 2 turboshaft engines, each developing 900 shp. These gave it a maximum continuous cruising speed at sea level of 161 mph, a range (with reserves) of 336 miles, and a hovering ceiling (out of ground effect) of 10,600 feet. Maximum takeoff weight was 9,500 lbs.

Truly versatile, the Army Lynx is usually referred to as a general purpose and utility helicopter. It can perform such roles as tactical troop transport (capable of carrying up to 10 fully equipped troops within its fuselage in addition to the normal flight crew of two), logistic support, armed escort of other troop-carrying helicopters, casevac, reconnaissance, search and rescue (SAR), and antitank. While other armament options are available, the principal weapon system carried by this version of the Lynx is the ubiquitous GM-Hughes BGM-71 TOW antitank missile. Eight of these highly effective weapons are carried, four on each side of the helicopter, with eight spare rounds stored inside the aircraft.

The standard TOW M65 target acquisition and missile guidance unit employed in conjunction with this weapon is produced in a roof-mounted version for the Army Lynx under license by a consortium led by British Aerospace. The weapons operator sits in the left seat of the Lynx to operate this sight and the related aiming and weapons management controls; the pilot sits in the right seat.

The current Army Lynx version is the AH Mk 7, which includes a more powerful Gem 41-1 engine that delivers 1,120 shp (the twin-engine layout is retained, allowing a maximum takeoff weight up to 10,750 lbs). It has a tail rotor that revolves in a direction opposite that of earlier versions. This results in lower noise levels and gives the helicopter greater ability to hover for long periods, both features particularly useful in the antitank role.

In adddition to its landing gear arrangement, the Navy Lynx differs from the Army Lynx in several other major respects. It contains a nose-mounted Ferranti ARI 5979 Seaspray radar, which is capable of detecting even small craft in high sea conditions. A TV screen within the helicopter gives all relevant data from this radar, including details of target range and bearing. It can be used in conjunction with British Aerospace SEA SKUA antiship missiles; up to four of these weapons are carried by the helicopter. In Royal Navy service, the Lynx carries the designations HAS Mk 2 and HAS Mk 3. The difference in designation reflects changes, including those of the powerplant — the Mk 2 having Gem 2 engines and the Mk 3 Gem 41-1 engines.

The Navy Lynx is fully capable of operating from ships. To ensure safe landings, particularly when operating from small stern platforms, a special hydraulically- powered deck lock is installed in the bottom of the helicopter's fuselage. During the landing, this connects with and grasps a metal lattice structure set into the deck. The helicopter is also held down securely on the deck until takeoff, when the lock is released. With an overall length (excluding the main rotor) of just under 40 feet (the Army and Navy Lynx are virtually the same length), the Lynx is a medium-sized helicopter. To facilitate stowage in cramped hangers and save space aboard ship, the entire tail section can be folded.

For submarine detection, the Navy Lynx can carry a Texas Instruments ASQ-81 magnetic anomaly detector (MAD). It can also be armed with British Aerospace Type 11 Mod 3 depth charges, or the Marconi Sting Ray antisubmarine homing torpedo. The Westland company is proposing a Super Lynx for enhanced naval operations and greater capabilities in its assigned missions. In this version, uprated Gem engines would be installed, plus an MEL Super Searcher 360-degree scanning search radar and an advanced dunking sonar. Weapons would include SEA SKUA, Sting Ray torpedoes, or the Norwegian Kongsberg Vapenfabrikk PENGUIN antiship missile.

The Navy Lynx has found many export customers (unlike the Army Lynx). Those helicopters operated overseas perform a variety of roles including SAR, antiship operations and antisubmarine warfare. The French Lynx carries an Alcatel DUAV-4 dunking sonar. It is armed with Aerospatiale AS. 12 wire-guided missiles, together with an SFIM APX M 335 gyrostabilized sight on the left side cabin roof. Dutch-operated antisubmarine Lynx aircraft also use the Alcatel dunking sonar, with some being equipped with the Texas Instruments ASQ-81 MAD equipment. The West German Navy Lynx features a Bendix AQS-18 sonar for antisubmarine operations, while the Nigerian Navy Lynx is provided an RCA 500 Primus radar rather than the standard Seaspray.

The Lynx 3 dedicated antitank helicopter is different from the standard Army and Navy Lynx. A prototype of this version first flew on 14 June 1984. In addition to the antitank model, a naval version is being proposed with Marconi Sting Ray torpedoes listed among its possible armament. Power would be provided by Gem 60 engines, each giving a maximum continuous rating of 1,115 shp, and allowing for a maximum takeoff weight of 13,000 lbs. The Lynx 3's landing gear is the tricycle-wheeled type. It is, nevertheless, different from the tricycle undercarriage found on the Navy Lynx. The fuselage has been lengthened by nearly one foot, providing increased storage space for spare missile rounds, or enabling it to carry as many as 12 fully equipped troops. Mobile antitank teams could also be carried, together with their missiles and launchers. The standard armament for the Lynx 3's antitank mission would be TOW, Euromissile HOT, or Rockwell HELLFIRE. It would also include air-to-air missiles such as the General Dynamics STINGER and various rocket pods, machine guns and cannons of differing calibers. A Martin Marietta target acquisition and designation system could be fitted on the aircraft as well as a pilot's night vision system.

Missions such as casevac, combat SAR, and resupply and logistic support could also be flown. For the latter, the total load carrying capacity is some 5,000 lbs, of which up to 4,000 lbs can be suspended from an external cargo hook. At present, no firm orders have been placed for the advanced Lynx 3, although this helicopter would be a very useful asset on the battlefield.

Another helicopter proposed with features of the Lynx is the TT 300 tactical transport. This helicopter will feature a completely new fuselage, 46 feet 8 inches in length (excluding the main rotor), with space for about 14 fully equipped troops. It could perform logistic support, SAR, and casevac missions. Power would be provided by two General Electric T 700-801 turboshaft engines, with a takeoff rating of 1,694 shp. A tricycle-wheel undercarriage is standard. Thus far, no firm orders have been announced for this proposed helicopter.

An important feature of new Westland projects, about which details have been revealed, is the role of special rotor blades. This is also becoming increasingly important on existing models. Westland is at the forefront in rotor blade technology, using advanced materials in various areas. This is well illustrated in the success of the Lynx demonstrator, G-Lynx, which broke the world helicopter speed record on 11 August 1986. Fitted with special rotor blades incorporating features from the British Experimental Rotor Program, the helicopter attained a speed of 249.1 mph. The record could not have been so easily broken without these features.

Another Westland product is the long-standing Sea King helicopter. This significant aircraft, in service with British forces and export customers, performs several vital roles. One is airborne early warning, which arose in large part from combat experience during the Falkland Islands War in 1982. This version carries a Thorn-EMI Searchwater radar and is the vital airborne early warning cover for the British fleet. Other Sea King versions can carry a variety of weapons including the Marconi Sting Ray homing torpedo and Aerospatiale AM.39 EXOCET antiship missile (as fitted first to Pakistani Navy models). In a recent development, West German Navy Sea Kings might possibly carry the MBB KORMORAN antiship missile in conjunction with the Ferranti Seaspray radar.

An Advanced Sea King also exists for service with British and Indian forces. With the latter, an MEL Super Searcher radar is used in conjunction with the British Aerospace SEA EAGLE antiship missile. This capable long- range weapon is one of several identified by Westland for use by the Advanced Sea King. Others include the McDonnell Douglas HARPOON, the SEA SKUA, EXOCET, and PENGUIN. Like the Lynx, the Sea King is set to remain in service for some time.

With the recent inclusion of the Sikorsky UH-60 Black Hawk in the Westland group as the WS 70 derivative, the quality and versatility of this proven line of helicopters have been greatly increased. Indeed, this list of very capable helicopters will be further extended when the Anglo- Italian EH 101 (a Westland-Agusta joint venture) reaches production. The Lynx, in particular, has forged an excellent service record, including valuable service with the Sea King during the Falkland Islands War. Today it is one of the West's most important multi-role helicopters. ⃝DI

aerospatiale **PANTHER**

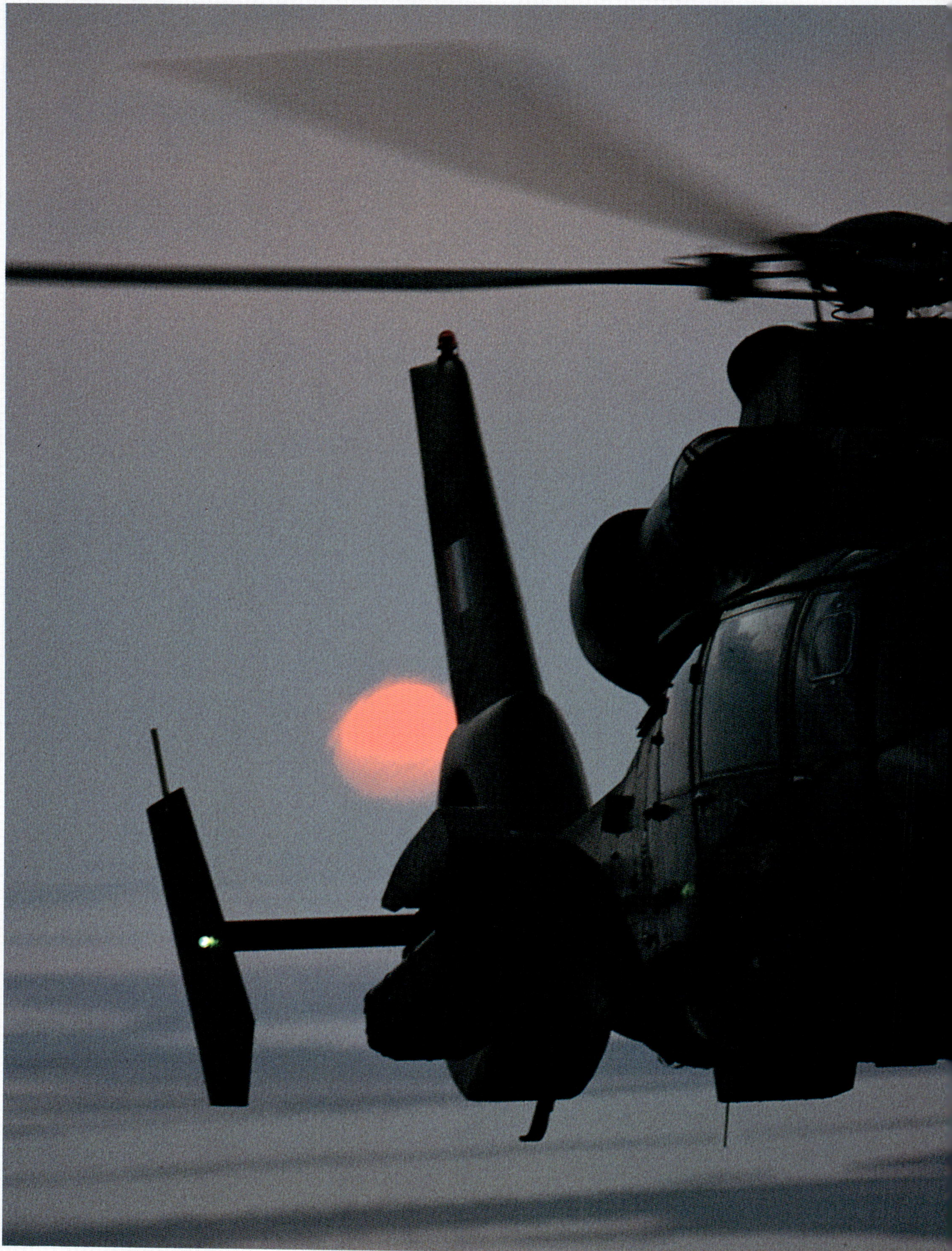

Aerospatiale SA 365 M Panther
by
Malcolm V. Lowe

One of the most successful helicopter designs that has been developed in recent years by the major French aerospace company, Aerospatiale, has proven to be the Dauphin (Dolphin) series of multi-purpose helicopters. With sales in the French domestic and foreign export markets, and even licence production in Communist China, the Dauphin 2, in particular, has become firmly established worldwide as a significant civil and military aircraft. It is a worthy hangermate of other highly successful Aerospatiale helicopters, which include such models as: the Ecureuil (Squirrel)/Astar; the Gazelle and the Puma which, together with the Lynx, were generally products of the Anglo-French helicopter agreement dating back to the second half of the 1960s; and the Super Puma transport and antishipping derivative of the basic Puma. Proving that a good basic design carries with it the potential for further development and adaptation, the basic twin-engine Dauphin 2 has been reworked by Aerospatiale to create a specific military version capable of performing a diversity of roles. This has resulted in another important Western European combat helicopter, the SA 365 M Panther.

The Aerospatiale company was formed in 1970. It quickly became, and has continued to be, a primary force among Western European helicopter manufacturers, enjoying an important hold on a significant part of the world helicopter market. The original Dauphin helicopter dates back to the early 1970s. In its initial layout, designated SA 360, the aircraft was a single-engine design with tailwheel landing gear. The first flight was made in 1972. In early 1973, the company announced a twin-engine development, designated SA 365. The first flight of this new model took place on January 24, 1975. This helicopter has subsequently led to all the current twin-engine, tricycle undercarriage production versions, which are known as Dauphin 2. Indeed, with the success of the initial SA 365s, the military potential of this new helicopter was soon realized.

55

In addition to many civil roles, this type serves with a number of air arms around the world.

Of particular note are the 20 military Dauphin 2s ordered by Saudi Arabia as part of a large arms deal. These helicopters, basically SA 365 F models, are powered by two Turbomeca Arriel 1M turboshaft engines, rated at 700 shp takeoff power and at 624 shp maximum continuous power each. The aircraft is optimized for naval antishipping missions. For this role it carries four Aerospatiale AS 15 TT command-guided missiles, which have a range of over nine miles, and a Thomson-CSF Agrion 15 radar for maritime surveillance and missile guidance. This advanced radar is housed in an externally-mounted flat circular radome beneath the helicopter's nose, giving it a 360-degree field of sweep. It has a track-while-scan capability, enabling it to detect possible threats while tracking some 10 targets simultaneously. The potential hit probability of this missile/radar combination is claimed to be very high, even against small targets, in adverse weather, or in the face of electronic jamming. The helicopter's standard flight crew is two, but up to 10 passengers can also be carried. Normal operational equipment can include Crouzet magnetic anomaly detection gear.

A dedicated antisubmarine warfare SA 365 F is also available, with all relevant equipment including provision for Alcatel HS 12 sonar. In addition to the Dauphin 2s, the Saudi Arabians ordered four SA 365 Fs fitted out for search and rescue (SAR) duties, equipped with an OMERA ORB 32 radar. All the Saudi aircraft, the first of which flew in 1982, can be operated from shore bases and the decks of frigates.

Also of great importance are the SA 366 G helicopters operated by the United States Coast Guard. These aircraft, designated HH-65A and named Dolphin in Coast Guard service, are different in several respects from standard versions of the Dauphin 2; much of their equipment and engines are U.S. manufactured. They were chosen for Coast Guard service in the face of strong competition to meet a requirement for a Short Range Recovery (SRR) helicopter, capable of operating from Coast Guard cutters and icebreakers, as well as from shore bases. The SA 366 G made its maiden flight on July 23, 1980, and initial deliveries to the Coast Guard began in 1984.

The operational SA 366 G-1 helicopters are powered by two Avco Lycoming LTS 101-750A-1 turboshaft engines, rated at 680 shp each. These give the Dolphin a useful maximum speed at gross weight of around 200 mph and a maximum cruising speed of 160 mph at sea level (again at gross weight), with an endurance of four hours on maximum fuel. Operational equipment includes full and advanced communications and all-weather search and navigation equipment, for which Rockwell Collins has a major responsibility. Of future importance for the Dolphin fleet is a Northrop Sea Hawk FLIR (forward looking infrared) for rescue operations in bad weather and low visibility, or even in the dark. Special inflatable flotation bags can be carried. The Dolphin's normal three-man flight crew consists of the pilot, co-pilot, and aircrewman/rescue hoist operator. The Coast Guard has a requirement for just less than 100 of these very capable aircraft.

The latest and most heavily armed military development of the Dauphin 2 series is the multi-role SA 365 M, which bears the name Panther. This combat aircraft made its first flight on February 29, 1984, and has since evolved into a well armed and equipped machine.

Exhibiting multi-mission capabilities, this new helicopter has been identified by Aerospatiale as having several primary roles, as well as a number of secondary tasks. To fulfill these diverse requirements, the Panther retains the basic layout of the Dauphin 2 helicopter, in particular the SA 365 N version, but with several important improvements. To begin with, the entire airframe is manufactured for high levels of survivability and crashworthiness.

The pilot and weapons operator sit in crashworthy, armor-plated seats. The helicopter's fuel tanks are self-sealing and there is additional protection for some of the fuel lines. The fuel tanks are capable of withstanding a crash with a dropping speed of up to 46 feet per second. Indeed, the whole fuselage structure can withstand an impact at a vertical speed (at maximum takeoff weight) of up to 23 feet per second.

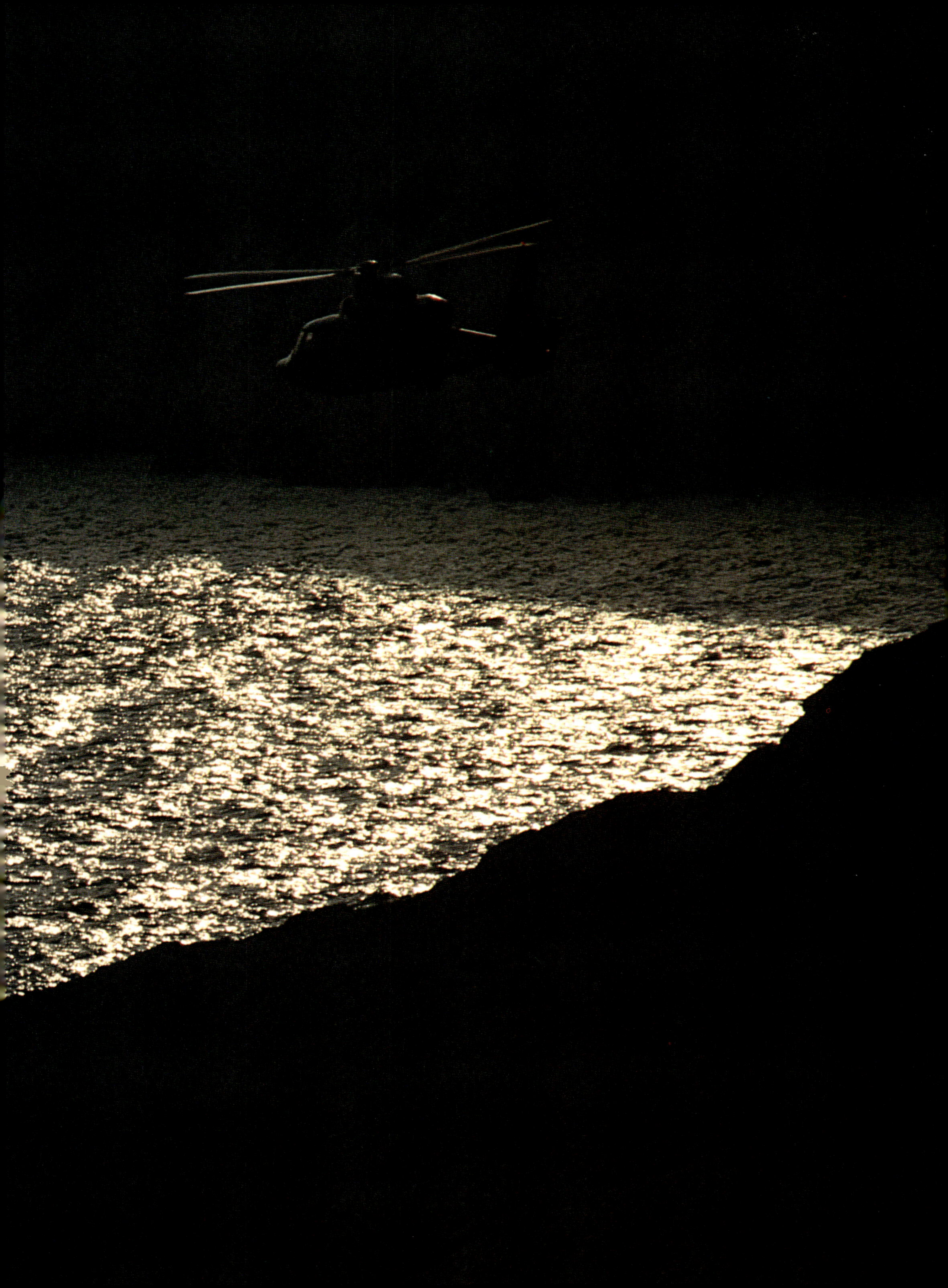

There is a large proportion of strong composite material included in the basic structure of the helicopter. This also extends to the Starflex fiberglass main rotor head, which successfully dispenses with conventional hinges by use of a single steel/rubber sandwich balljoint for each main rotor blade, and to the main rotor blades themselves. These blades, constructed principally from fiberglass and various composites, in similar fashion to those fitted to standard Dauphin 2 helicopters, can survive hits from .5-inch ammunition. A carbonfiber fenestron with enclosed tail rotor is incorporated, the blades of which can withstand small caliber bullet impacts. The fenestron is a standard feature of the Dauphin 2 helicopter, but is featured in a slightly modified form on the Panther.

The helicopter's composite materials help to reduce the overall radar signature, a desirable effect which can be enhanced through the use of special paints. Infrared (IR) signature is considerably reduced by such features as a jet dilution device fitted behind the engines, which dissipates partially cooled exhaust gas upwards into the main rotor downwash. Enveloping engine fairings and the use of low IR-reflecting paint also reduce the overall IR signature. Attention has been given to reducing noise levels from the helicopter. To further reduce vulnerability, redundant hydraulic and electrical circuits are included. The helicopter can carry a range of electronic countermeasures and related support equipment, including chaff dispenser, IR jammer, and a Sherloc radar warning sensor.

The Panther is powered by two Turbomeca TM 333-1M turboshaft engines, each producing 912 shp maximum contingency power, 838 shp takeoff power, and 751 shp maximum continuous power. These engines give the helicopter a maximum speed at gross takeoff weight of 184 mph, and a maximum cruising speed at sea level (and at gross takeoff weight) of 172 mph. It has a hovering ceiling (out of ground effect) quoted at 7,546 feet, and a range with standard fuel tanks of over 460 miles. The gross takeoff weight is 9,039 lbs. It measures just short of 40 feet (excluding the main rotor), with a maximum fuselage width (main cabin) of some six and a half feet. A fully retractable tricycle-wheeled undercarriage is installed.

Among the primary roles identified by Aerospatiale for the Panther is troop transport. Towards this mission, the Panther's relatively capacious main cabin is able to accommodate up to 10 troops (in addition to the helicopter's two crew members) in a high density seating arrangement, or eight troops with slightly more room to spare. Main cabin access is facilitated by four doors, two on each side of the main cabin. The rear door on each side slides aft and the front door opens outward, hinged at its forward edge. The Panther's comparatively high speed and low vulnerability make it a very useful tactical troop transport. If required, the troop seats can be rapidly folded and stowed inside the helicopter's hold, which itself can also be used for carrying an extra fuel tank or equipment.

Normally, in the troop transport role, the Panther would be lightly armed or unarmed. For another possible primary mission, fire support, a full armament kit can be carried. The Panther can be fitted with a universal weapons beam for mounting a wide variety of munitions on each side of the fuselage. For fire support against enemy personnel or lightly armored vehicles, the Panther can carry on each side a Brandt rocket launcher with 22 68mm rockets, or a Forges de Zeebrugge pack of 19 2.75-inch rockets, or a GIAT M-621 20mm cannon pod with 180 rounds of ammunition. For accurate firing of the helicopter's armament, the pilot is equipped with a Crouzet electronic sight and aiming system. The weapons sight is mounted in the pilot's line of vision and is capable of presenting relevant symbology for accurate aiming. The pilot also receives flight handling assistance through the helicopter's SFIM 155D autopilot. A useful function of the flight handling assistance is the automatic negation of the tendency for the helicopter's nose to drop when the cannons are being fired. This dropping of the nose is caused by the turning moment of the weapon's recoil about the helicopter's center of gravity, which rotates the nose downward if not prevented.

The Crouzet sighting system in a modified form can allow for air-to-air firing (together with the installation of an air-to-air firing computer).

Another primary role suggested by Aerospatiale for the Panther is helicopter escort and protection against enemy aircraft. In this role, the Panther could escort transport and antitank helicopters, either carrying the 20mm GIAT cannons, or up to eight Matra MISTRAL IR homing air-to-air missiles.

Of great importance is the task of day and night antitank warfare adopted for the Panther. For this mission, the helicopter can carry up to eight Euromissile HOT antitank missiles. Full night, as well as daytime, capability is provided by the installation of a Viviane sight system and related equipment fitted on the left side of the helicopter's cabin roof and operated from the left cockpit seat. For target detection, Viviane has an optical channel for use during daytime and an IR channel for night work, while a laser rangefinder provides distance-to-target and related measuring. The HOT system can be operated in much the same way by day or by night. The helicopter's crew can be equipped with full night vision goggles. This night antitank capability is one of the Panther's most significant roles. The overall endurance for antitank missions can reach around three hours. The Panther's suitability for this and other combat roles is greatly enhanced by its considerable stability as a weapons platform and by its maneuverability and capability to perform nap-of-the-earth (NOE) flight.

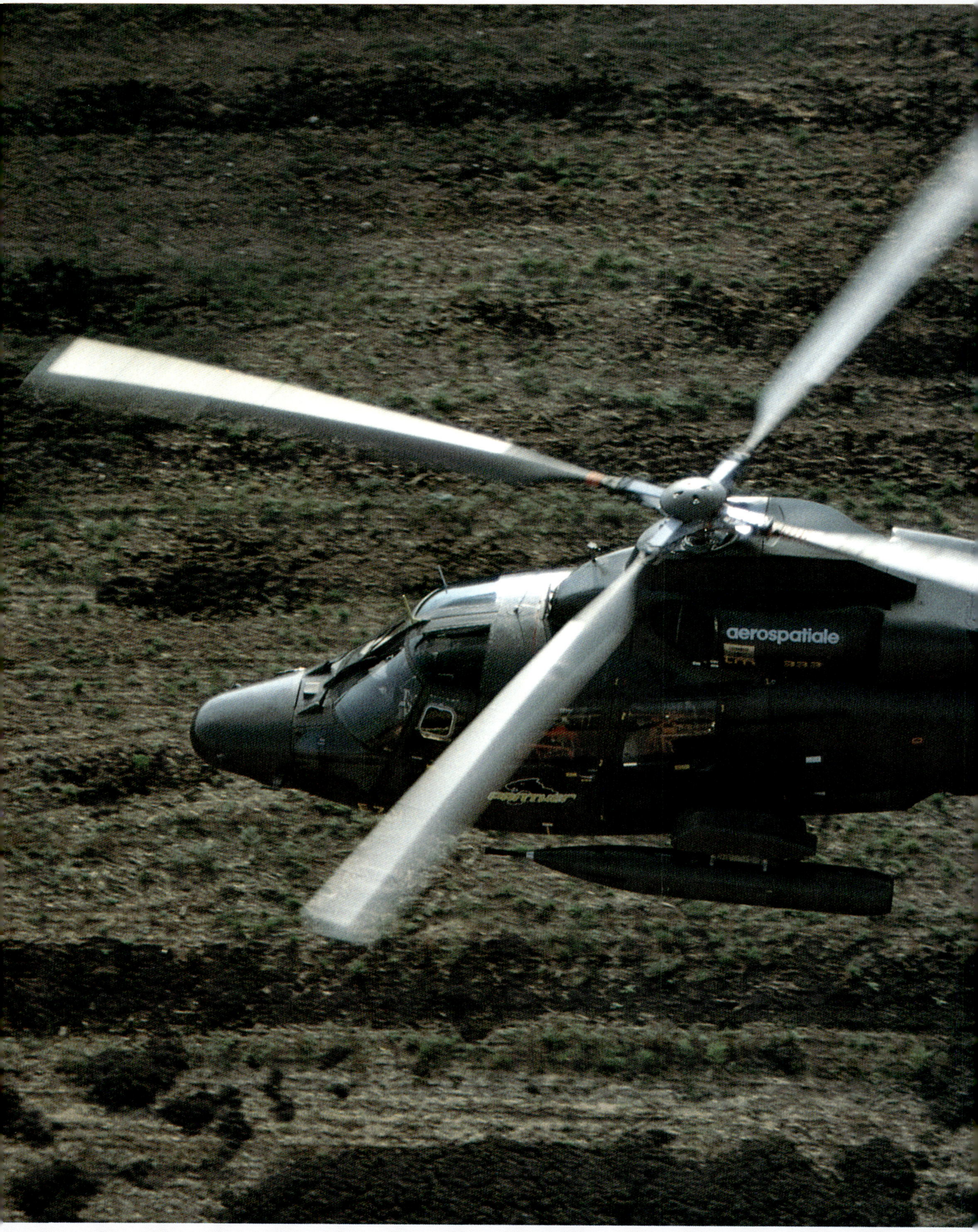
aerospatiale

The Viviane sight can also be employed for use on armed reconnaissance missions, in conjunction with the helicopter carrying weapons suited to this particular primary role, such as the MISTRAL missile or 20mm cannons. Armed reconnaissance is the final primary role examined in some detail for the Panther. The various weapons options available to the Panther can be carried in a variety of combinations, with an armament quick-change capability being another of the helicopter's features.

Aerospatiale has identified several secondary roles that the Panther can perform with considerable ability. These include: (1) casualty evacuation, for which up to four stretchers plus an attendant or sitting patient can be carried in the helicopter's main cabin; (2) combat SAR, with provision for the installation of an external winch on the helicopter's right side (the Panther can carry an extra crew member as winch operator and up to three rescued people while loaded with weapons on the left side of the fuselage in their normal position); (3) logistical support, with the Panther's main cabin being easily converted to carry large single items, in addition to the capability for installation of an external cargo sling beneath the helicopter; and (4) passenger/VIP transport, for which several possible internal seating arrangements can be installed, including a customized version featuring a deluxe VIP interior with enhanced sound proofing and air conditioning.

A production-standard Panther has already been fully demonstrated. Production Panthers are scheduled to be available from 1988, with the capability of flying many of the roles already envisaged. With its extensive range of possible missions and its ability to fly at altitude and in hot climates (special optional equipment available includes sand-prevention filters for desert operation), the Panther is a versatile and capable helicopter. Derived from an existing design, the fully combat capable Panther has inherited many of its best features and combined them with some exceptional characteristics, making the Panther a truly multi-role helicopter — one that should find many ready customers to add to Aerospatiale's proven sales successes around the world. IDI

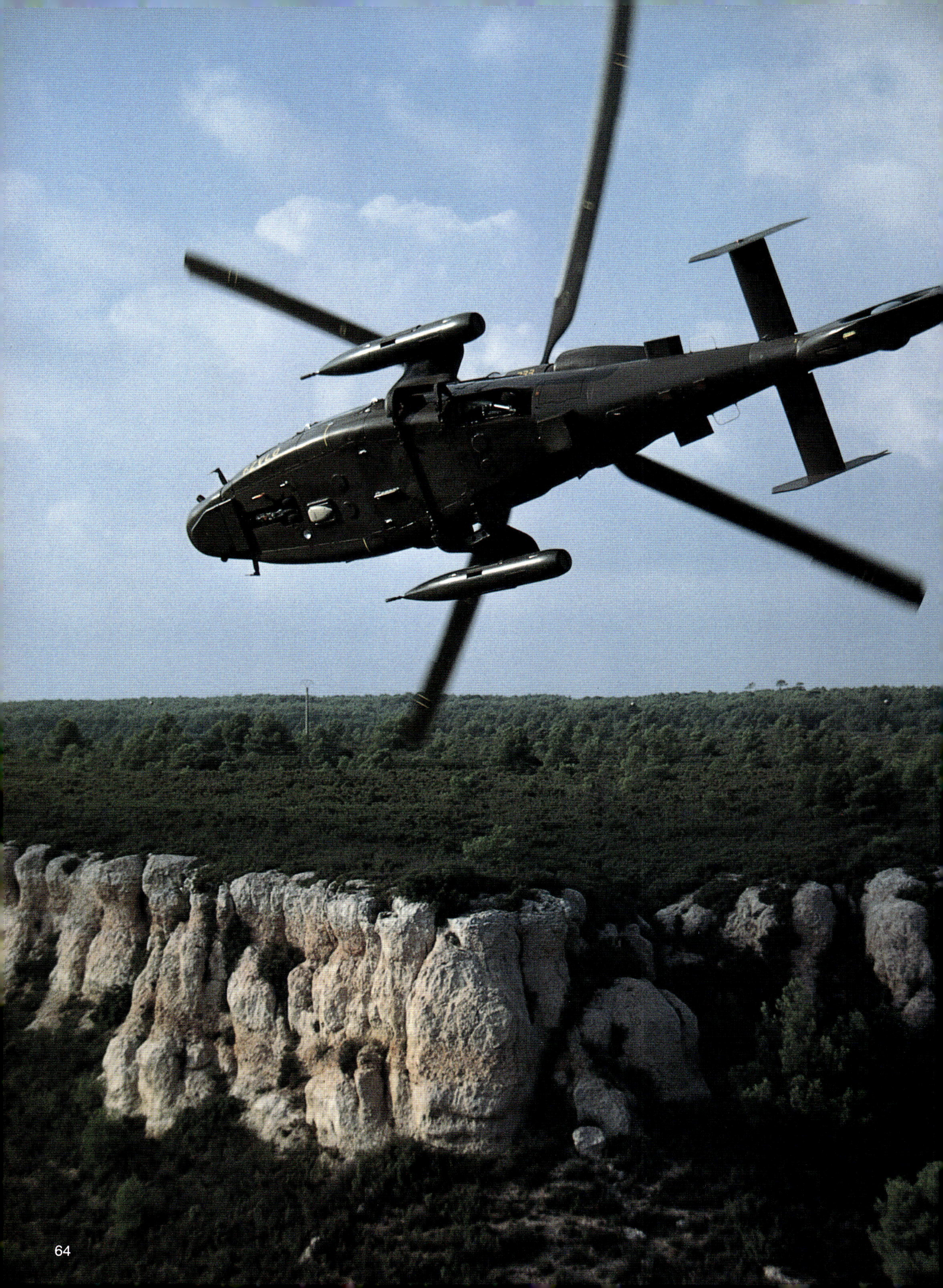

MBB BO 105/PAH-1
by Malcolm V. Lowe

The Messerschmitt-Bolkow-Blohm (MBB) BO 105, with several years of military service as well as accomplished employment in civil, paramilitary, and police service, has become firmly established as one of Western Europe's most important helicopter designs. It is considered to be among the most capable light helicopters yet produced anywhere in the world. Indeed, the whole BO 105 program has grown into one of West Germany's most successful military aircraft projects. This is attested to by MBB's sales of approximately 1,000 BO 105s of all types to 34 countries on four continents. Of these, over 600 have been supplied to some 18 military users worldwide. Included are aircraft produced in West Germany as well as those manufactured under license overseas. A significant number are armed helicopters dedicated to the antitank role.

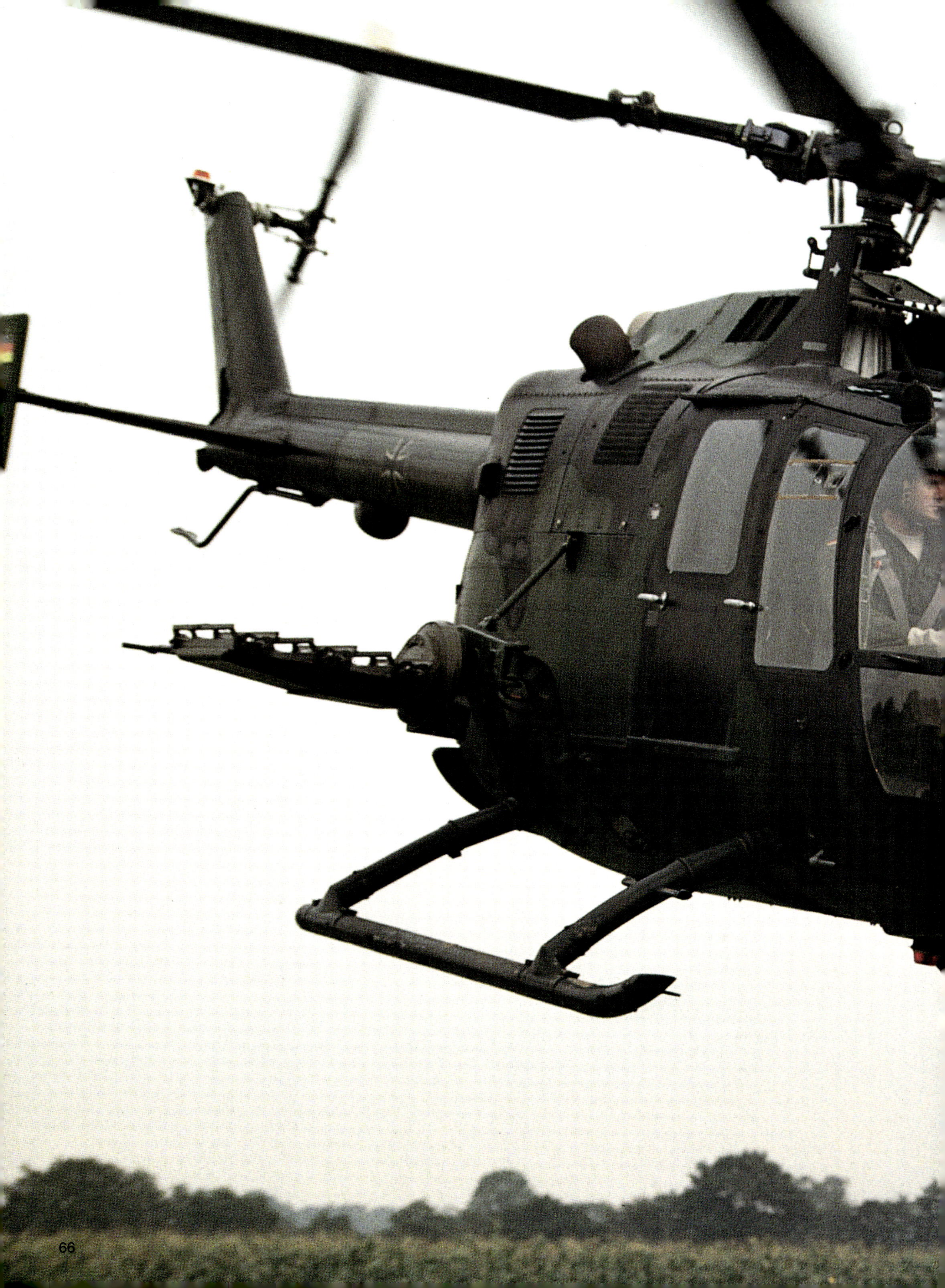

Even before the MBB group was formed in 1969, the independent Bolkow company enjoyed years of experience in helicopter design and research. As originally conceived in the 1960s, the BO 105 contained several innovations that were to become significant features of the aircraft subsequently produced. Although small and comparatively light from the start, this helicopter featured the innovation of a twin-engine design with the safety factor of single-engine operation. Other novel features installed included Bolkow's own hingeless rotor system.

Under West German government sponsorship, the BO 105 project proceeded in 1964 with the construction of three prototypes, two years after design work had begun in earnest. The first flight was made by the second prototype on 16 February 1967. As a result of tests and evaluation, series production began in the early 1970s. The military potential of the new helicopter was soon recognized. In 1974, after severe evaluation programs, the BO 105 was selected by the West German Ministry of Defense to fulfill a light liaison/observation helicopter requirement for Army aviation (Heeresflieger). The version produced for this role was designated BO 105M or VBH. A year later, the BO 105 was selected to fulfill a different requirement for Heeresflieger service — as an antitank helicopter able to carry antitank missiles. This version was designated BO 105P or PAH-1, the latter standing for Panzerabwehrhubschrauber der 1 generation, or antitank helicopter of the 1st generation.

Full development of the PAH-1 led to production of 212 aircraft for the Heeresflieger, with delivery completed during 1984. Apart from its armament and related equipment, the PAH-1 is similar to the unarmed BO 105M (of which 100 were delivered to the Heeresflieger). Both versions are powered by two Allison 250-C20B turboshaft engines, each producing a takeoff power of 420 shp and a maximum continuous power rating of 400 shp. Additionally, both versions (in comparison to standard civil models) feature a strengthened transmission system, more efficient tail rotor, reinforced rotor components, crash-proof and rupture-proof fuel system, and skid landing gear able to absorb much rougher landings.

Servicing has been simplified to allow more time between overhauls and to permit a high level of maintenance in the field. Under wartime conditions, these helicopters generally would have to operate away from established air bases. Modular construction allows defective systems to be removed and replaced easily and quickly. Maintenance concept parts are only removed when and if necessary. Similarly, the number of special tools and test facilities required has been minimized.

Slightly larger than the McDonnell Douglas Helicopter (formerly Hughes) Model 500/530 Defender series, the PAH-1 is still a small helicopter. It measures just over 28 feet in length (excluding the main rotor), and has a main rotor diameter of slightly more than 32 feet 3 inches. This small size increases the helicopter's ability to operate in the vicinity of trees and also makes it difficult to be spotted visually, particularly when combined with a dark camouflage scheme. The PAH-1 features considerable maneuverability, even at low level, which allows it to perform nap-of-the-earth (NOE) flight using natural terrain for concealment. A key to this maneuverability is Bolkow's hingeless rotor system, which eliminates the flap and lag hinges required in conventional rotor systems employing metal rotor blades. This is achieved through the use of fiberglass-reinforced plastics for the rotors. The inherent elasticity of this product absorbs the flap and lag movements and the stresses imposed on rotor blades in flight. MBB states that such a system, which allows superior maneuverability and excellent response, is much better than that of most rotor designs. It also helps to reduce vibration while additionally easing maintenance due to the fewer number of moving parts required. A rigid titanium rotor head is fitted.

The PAH-1 has a maximum speed (at gross weight) of 150 mph, and a range at sea level (with no reserves) of slightly less than 200 miles. Endurance is just over two hours; the helicopter has a hovering ceiling (out of ground effect) of 5,200 feet. It weighs 3,688 lbs empty, and has a maximum takeoff weight of 5,291 lbs, allowing it to carry full missile armament. In West German service the PAH-1 has a standard weapons fit of Euromissile HOT (high subsonic, optically-guided, tube-launched) antitank missiles, plus their associated sighting and related equipment.

The HOT missile is an effective antitank weapon. Each missile contains a hollow-charge warhead weighing 13.2 lbs and is powerful enough to penetrate existing conventional armor. It is also effective against newer composite armor. The range of the missile is from 1,310 to 13,125 feet. It takes approximately 17 seconds to reach a target at the maximum range with a velocity that is almost constant at 787 feet per second. At maximum distance from the target, the helicopter is theoretically beyond the range of enemy guns and light surface-to-air missiles. The PAH-1 can carry six HOT missiles, three on each side. They are mounted on their own launch ramps on a special pylon within fiberglass tubular containers.

Some structural reinforcement was required to allow the helicopter to carry HOT missile equipment, but the overall system was fully installed with little difficulty. The missile has semi-automatic command to line-of-sight wire-guidance, using an SFIM APX M 397 gyrostabilized sight system and related equipment mounted on the left side of the helicopter's cabin roof. The weapons operator, who sits in the helicopter's left seat, is provided the necessary controls for sighting, aiming, selecting, and firing the missiles, plus all related functions. With placement of the sighting system on the cabin roof, the helicopter can hover almost concealed behind the tops of trees or other suitable cover. Since only the extreme top of the fuselage and rotor are visible, the helicopter cannot easily be spotted by the enemy while it searches for, tracks, and engages targets. The HOT system is capable of comparatively high rate of fire. The missile's hit probability of 90 percent makes it a valuable asset in the variety of weapons available to NATO forces.

In addition to the Heeresflieger-operated, HOT-armed PAH-1, 28 similar aircraft (designated BO 105 ATH) operate with Spain's Army aviation. These helicopters have been license- manufactured in Spain by Construcciones Aeronauticas SA (CASA). The Spanish Government viewed the PAH-1 as a cost-effective, capable antitank helicopter that meets their requirements. The BO 105 ATH is joined by 18 unique BO 105GSHs armed with a Rheinmetall Mk 20 Rh202 20mm cannon mounted below the helicopter's fuselage.

The basic BO 105 design has proven quite versatile with weapons options. A Multi-Purpose Delivery System (MPDS) has been developed for the type that includes detachable multi-purpose pylons and related internal equipment. This permits a wide range of armament and stores to be carried, including unguided rockets of various calibers, machine gun pods, cannon, reconnaissance equipment, and forward looking infrared pods.

Naval operations have been fully explored by MBB, for both antiship and antisubmarine warfare (ASW) roles, as the BO 105 is capable of operating from the decks of ships. To perform the ASW role, a magnetic anomaly detector can be carried, plus one or two lightweight ASW torpedoes such as the Whitehead Motofides A.224/S, developed for the Italian Navy.

License production of the PAH-1 helicopter by CASA has raised the possibility of several export models being delivered to Iraq. Sweden is also an export customer for antitank BO 105s. In this case, however, the 20 helicopters involved are not the HOT-equipped PAH-1s in service with West Germany and Spain. They are close to standard BO 105CBs, configured to carry SAAB/Emerson HeliTOW antitank missiles. This illustrates the compatibility of the BO 105 and PAH-1 with these missiles and the standard GM-Hughes BGM-71 TOW as well, for which a certification program already exists. The Swedish aircraft carry a sophisticated SAAB Helios sight, providing night vision and laser capability for night attack missions.

It is possible that a full night operational capability will eventually be incorporated in West German PAH-1s. A day/night gyrostabilized sight known as Viviane has already undergone testing, together with features such as night vision goggles for the crew. MBB has proposed a major upgrading of the PAH-1 fleet, beginning around 1990. This would include such improvements as: installation of Allison 250- C20R-3 engines; a higher maximum takeoff weight; advanced rotor blades; a lighter fire control system (the current equipment is based upon that fitted in the Leopard 1 main battle tank); and Euromissile HOT-2 armament. Various projects exist which could further enhance the basic PAH-1's capability. These include mast-mounted sights such as the SFIM-developed Ophelia system, which gives an unrestricted 360-degree view around the helicopter. This system can be adapted to the existing helicopter with a minimum of structural modification. Nose-mounted equipment could include the Pilot's Infrared Sighting Ability (PISA) system developed by MBB's Dynamics Division. This provides a night vision capability for orientation and observation purposes. Martin Marietta's Pilot's Night Vision Sensor (PNVS), which can display via a helmet-mounted system, is another possibility. Such advanced equipment allows safe flying, even at night or in very poor visibility. Advanced cockpit layouts, with such features as multi-function cathode ray tube displays of the very latest technology, are also under study.

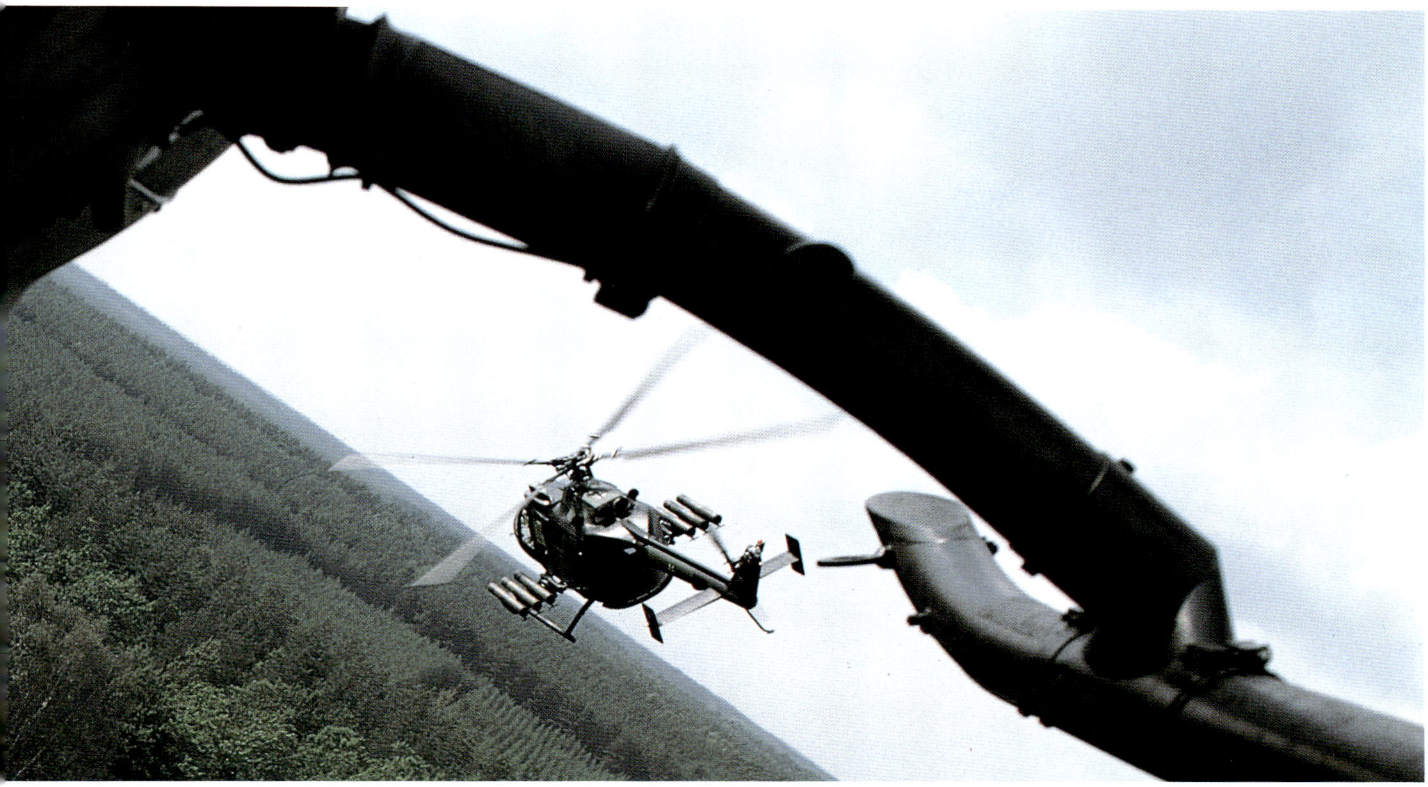

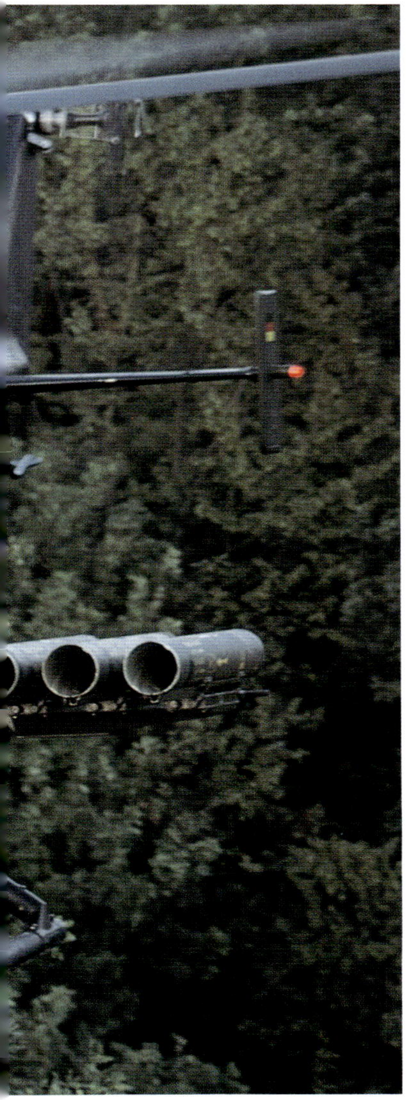

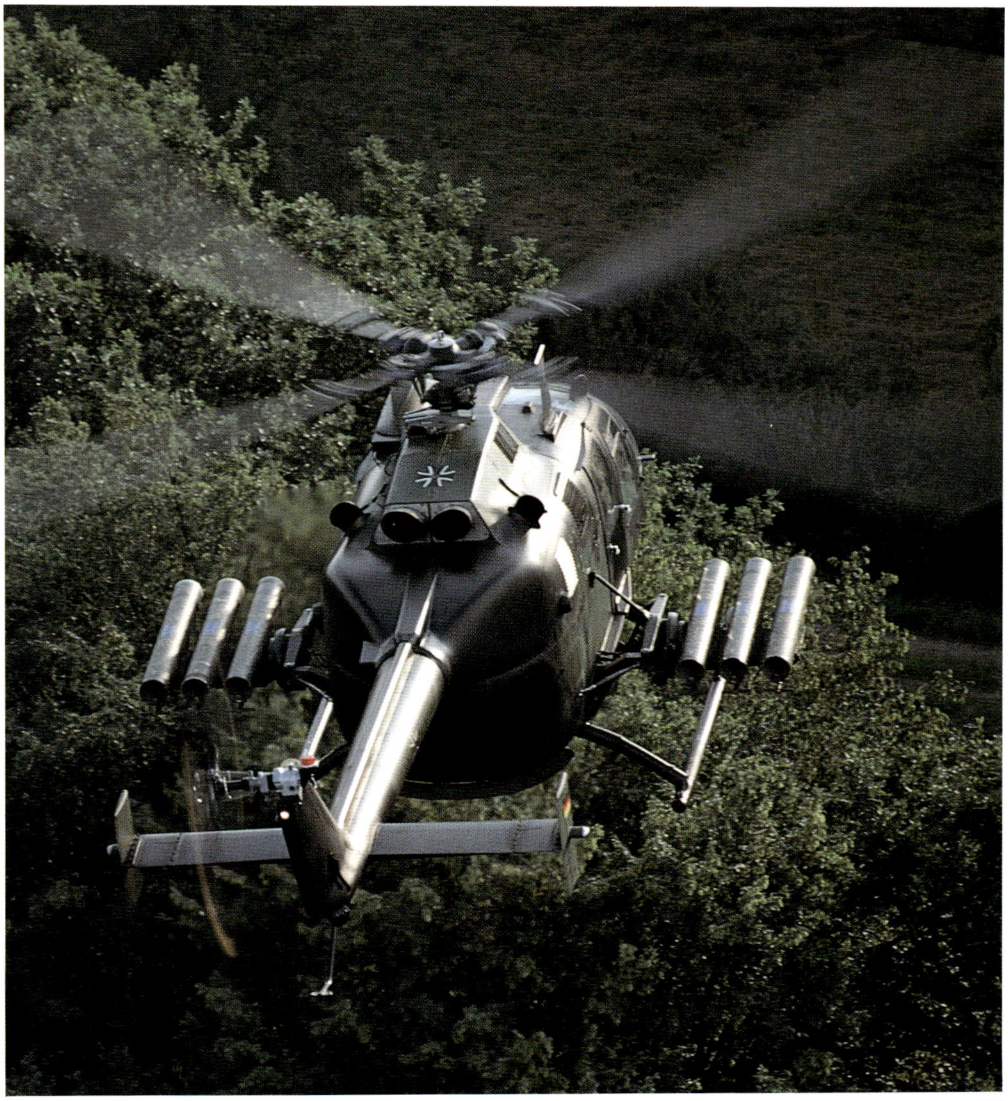

The second generation PAH helicopter for the Heeresflieger, the PAH-2, is currently under study. This helicopter will be quite different from the PAH-1 in that it is intended to serve as a dedicated antitank helicopter. It will be produced as a joint international venture between MBB and Aerospatiale of France. A Memorandum of Understanding covering the project was signed by France and West Germany in 1984 and a new organization, Eurocopter, was created to manage the project. West Germany wants 212 of the PAH-2s, which will resemble other dedicated antitank helicopters in design. In the years ahead, much work will be carried out on this project, which will include versions to serve with the French military forces. The PAH-2 will have a fine predecessor in the PAH-1, which has set such a very high standard of service for helicopters. IDI

AGUSTA A129 MONGOOSE
by Malcolm V. Lowe

The advanced and formidable Agusta A 129 Mangusta (Mongoose) is a helicopter with a very important claim to fame — to date it is the only dedicated all-weather, day/night antitank helicopter to reach production status in Western Europe. It is some years ahead of any potential rivals in Western Europe, and set to become one of the West's most important antiarmor helicopters.

The Agusta company, however, is not only producing an advanced antitank helicopter of outstanding capability. The antiarmor Mangusta is only one of several planned versions of the A 129 that represent a whole family of helicopters intended for diverse military roles.

Over many years, Agusta has become firmly established in helicopter design and manufacture, as well as in license production and modification of existing designs from Sikorsky, Boeing, McDonnell Douglas and Bell. Versions of the company's A 109 helicopter have become highly successful civil and military aircraft of considerable capability. Indeed, during the 1970s when the Italian Army and Ministry of Defense began formulating a need for a dedicated all-weather, day/night light antitank helicopter, Agusta examined a development of the basic A 109 as a real possibility in meeting the stringent requirements.

Full preliminary design work to meet the Italian military specifications began in 1978. By 1980 the A 109 layout had given way to an all-new design based on the latest available technology and innovations. Agusta's designers were able to incorporate the best features into the new design. During this phase, they proposed a helicopter with increased power and lethality to achieve the greatest possible effectiveness and capability. With the creation of the A 129 program, a number of other possible roles were examined. This resulted in a whole family of military helicopters being devised, with the various planned versions configured with similar engines, rotors, drive systems, and other major components.

Five flying prototypes were subsequently built, and the first Mangusta made its official maiden flight on 15 September 1983. The Mangusta thus became the first member of the planned overall A 129 family to reach flying status. Agusta states that the Mangusta program ran on schedule. At the beginning of 1986 work was commenced on the industrial production of the first batch of 15 Mangustas, with deliveries scheduled to begin before the end of 1987. A total of 60 of these aircraft have been ordered for service with Italian Army aviation squadrons. A requirement exists for 30 additional helicopters plus reserves to equip another squadron. In addition to Italy, several other countries have shown considerable interest in the Mangusta, particularly the Netherlands.

The Mangusta displays some of the classic features of the special antitank helicopter layout. The long, thin fuselage, excluding the main rotor, measures just over 40 feet. Its maximum width, excluding the helicopter's stub wings, is just over three feet, giving the aircraft a small frontal area that is difficult to spot. The two-man crew sits in tandem with the weapons operator in the front seat; the pilot is situated behind and above him. Both crew members have an excellent field of view, particularly forward, out of the large cockpit side windows. Each is provided a Martin-Baker HACS 1 armored crashworthy seat specially developed to satisfy the A 129 requirements. Many features have been incorporated to provide unsurpassed crashworthiness. It can withstand a vertical impact of 36.5 feet per second, and contains roll bars for crew protection as well as other crashworthy advantages. The fixed tailwheel undercarriage can withstand hard landings of up to 15 feet per second descent rates. The whole airframe features ballistic tolerance against .5-inch armor piercing ammunition; it also resists 23mm projectiles. The helicopter has self-sealing and crash-proof fuel tanks, and self-sealing fuel lines. Further, all relevant flight mechanical linkages and moving parts normally found around the rotor mast are housed within the Mangusta's mast for added protection and to reduce the radar signature.

Many systems are designed redundantly. For example, there are two separate fuel systems with crossfeed capability. The hydraulic system includes three principal circuits for the flight controls and two independent circuits for rotor and wheel braking. There are dual fly-by-wire systems as a backup for the mechanical control system, and a separate fly-by-wire control system for the tail rotor with mechanical backup. All of the helicopter's main functions are handled and monitored by an aircraft management system called the Integrated Multiflex System (IMS). Jointly developed by Agusta and the Harris Corporation in Florida, this innovative and highly important system is managed by two redundant central computers. It controls and manages, among many other functions, the helicopter's performance, flight controls, and electronics, and performs such roles as engine monitoring and fuel and hydraulic monitoring and control. Processed information is supplied to the crewmembers on head-down multi-function displays which have standard multi-function keyboards for access to specific information. Navigation is controlled by a navigation computer in the IMS coupled with a GEC Avionics doppler radar, air data system and radar altimeter. The IMS can also collect data and warn of any necessary maintenance of components.

To allow for full night as well as day operations, a nose-mounted Honeywell night vision system is installed with an associated FLIR (forward looking infrared). This presents a picture of the world outside via a monocle of the pilot's IHADSS (integrated helmet and display sighting system). It provides a true and totally usable head-up image. The weapons operator also has an IHADSS. The Mangusta, therefore, can fly nap-of-the-earth (NOE) with safety, even at night. For night antitank operations, the A 129 uses the Heli TOW A, developed by Emerson Electric, SAAB Scania, and Agusta, which houses a FLIR system.

Systems redundancy within the Mangusta extends to a twin-engine layout, the helicopter being powered by two Rolls Royce Mk 1004 turboshaft engines. These have a maximum continuous rating for normal twin-engine operation of 815 shaft horsepower (shp) each, and an emergency rating of 1,035 shp each. With this power, the Mangusta has a dash speed of around 196 mph and a maximum level speed at sea level of 161 mph. It has a maximum takeoff weight of 9,039 lbs and a hovering ceiling (out of ground effect) of 7,850 feet, when the temperature condition is 20 degrees Celsius (C) higher than the engineering standard temperature (ISA). The maximum endurance with no reserves is some three hours. Low engine noise levels and infrared engine exhaust suppression are included to further reduce the Mangusta's detectability. The four-bladed main rotor features extremely strong rotor blades constructed from composite and fiberglass materials with a stainless steel leading edge abrasion strip. It can withstand hits from .5-inch and 23mm ammunition. Low-noise blade tips are included. The tail rotors are also of composite structure with a stainless steel leading edge, and can take hits of .5-inch caliber.

    The Mangusta can carry a comprehensive fit of active and passive electronic countermeasures and electronic counter-countermeasures equipment. Passive systems can include a radar warning receiver and a laser warning receiver. Active countermeasures can include radar jammer and an infrared jammer. A chaff and flares dispenser can be carried.

    To fulfill its antitank role, the Mangusta is able to carry a wide range of weapons. The stub wing on each side of the fuselage has an inner pylon able to carry loads up to 660 lbs, and an outer pylon for loads up to 440 lbs. In a typical weapons combination, eight standard TOW antitank missiles can be carried on the outer pylons (four on each side) together with a pod under each inner pylon for seven 2.75-inch air-to-surface rockets or further TOW missiles. Alternatively, eight Euromissile HOT or Rockwell AGM-114A HELLFIRE antitank missiles can be carried, with combinations of .5-inch or 20mm gun pods, and various types of rocket pods, including those for 81mm rockets. In addition, the helicopter can defend itself against enemy air threats, a very useful capability in view of the increasing danger from Soviet armed helicopters which also include an air-to-air capability. Thus, either Matra MISTRAL or General Dynamics STINGER missiles can be carried beneath the standard weapons pylons, or an AIM-9 SIDEWINDER can be mounted on a launch rail fitted at the tip of the stub wing. Overall, the Mangusta can pack a very deadly punch. All of the various armament combinations can be managed by the IMS through its fire control subsystem.

Such a wide range of weapons options has allowed Agusta to identify future scout/close support and air defense roles for the A 129 apart from the standard antitank mission. The company has explored the possibility of installing a chin-mounted .5-inch machine gun or a suitable cannon. In addition, an antiship version has been proposed. This version would utilize the same basic long and thin fuselage as the current antitank Mangusta, but would be armed with British Aerospace SEA SKUA or Oto Melara Marte Mk 2 missiles. The SEA SKUA is a fully combat-proven antiship weapon, having been successfully employed by British forces during the 1982 Falkland Islands conflict.

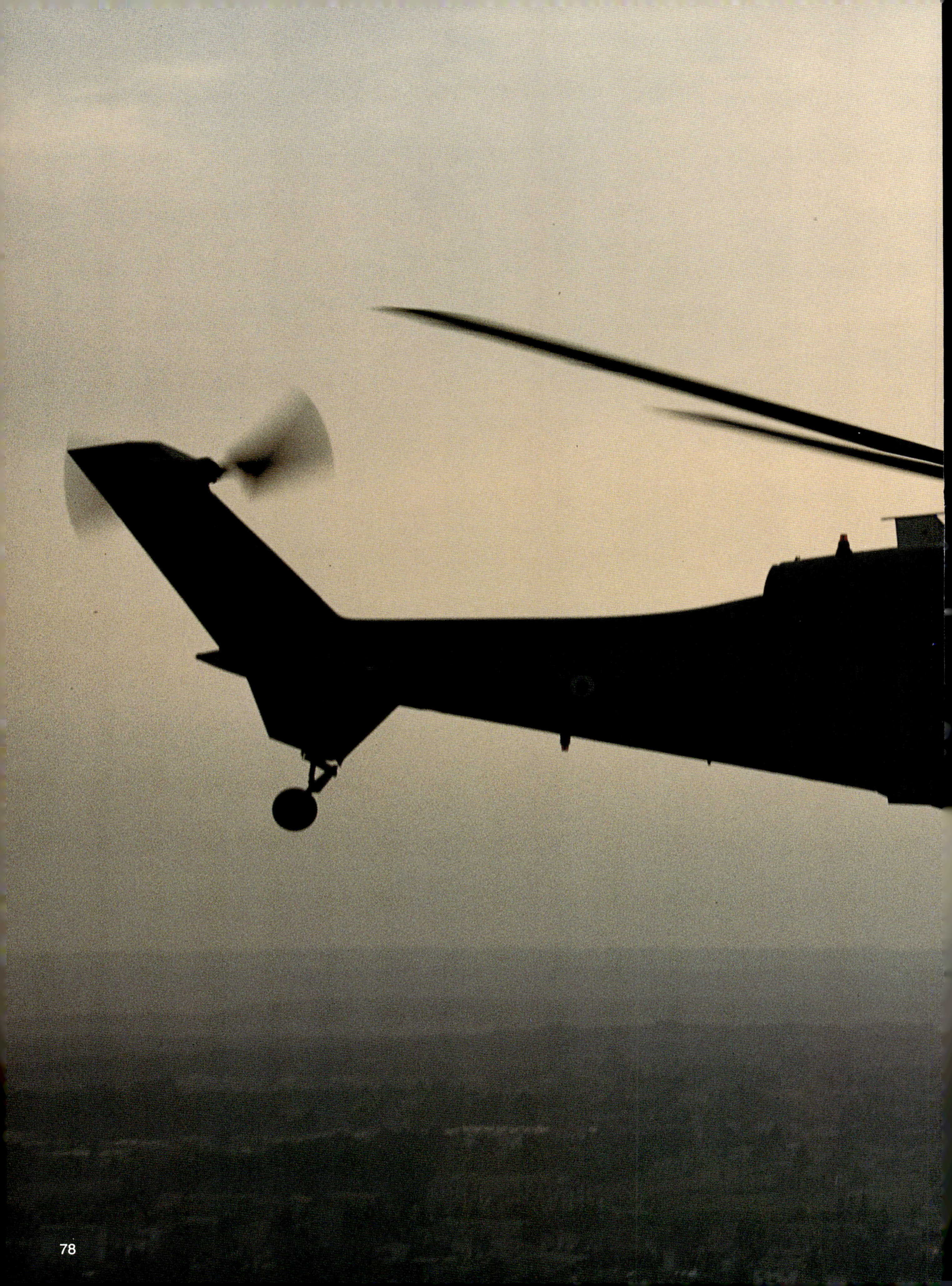

A further proposed member of the A 129 family is a light battlefield support helicopter. This would have a completely new forward fuselage with side-by-side seating for the two crew. The widened fuselage would provide a space for an assault squad of up to 10 soldiers. It would have many of the Mangusta's systems and survivability capabilities, and would feature several armament options. The aircraft would also be capable of flying combat rescue and casevac missions, and conducting SAR operations.

Based to an extent on the A 129 layout is an international helicopter program comprising Agusta, Fokker, Westland and CASA, aimed at creating a helicopter now named Tonal. This helicopter would be similar to the Mangusta and would perform antitank, scout, and antihelicopter roles using more efficient fly-by-wire systems, considerably more composites, and third generation weapons such as the TRIGAT antitank missile.

The A 129 Mangusta, with its excellent capabilities, performance, range of lethal armament options, simplified servicing through such factors as on-condition maintenance, survivability, and air-to-air capability, is an extremely important helicopter. It will give NATO antitank forces a massively increased capability when reaching full service in the near future. For years to come, this helicopter will be one of the West's most prominent weapons in any future conflict in Western Europe. A viable front line aircraft, it will reach full operational status long before any comparable helicopter produced in Western Europe enters service. IDI

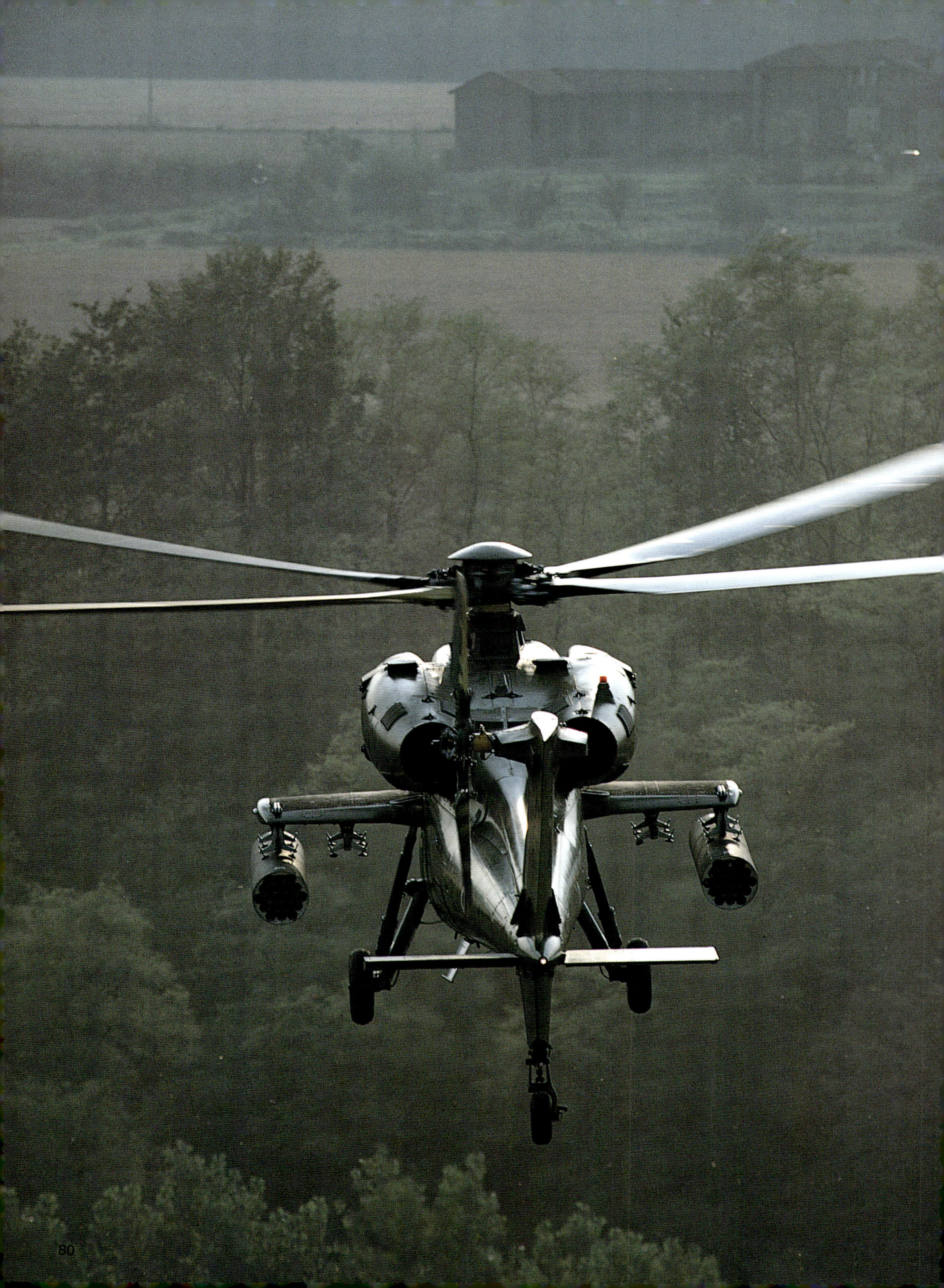

Modernized AH-1S Cobra

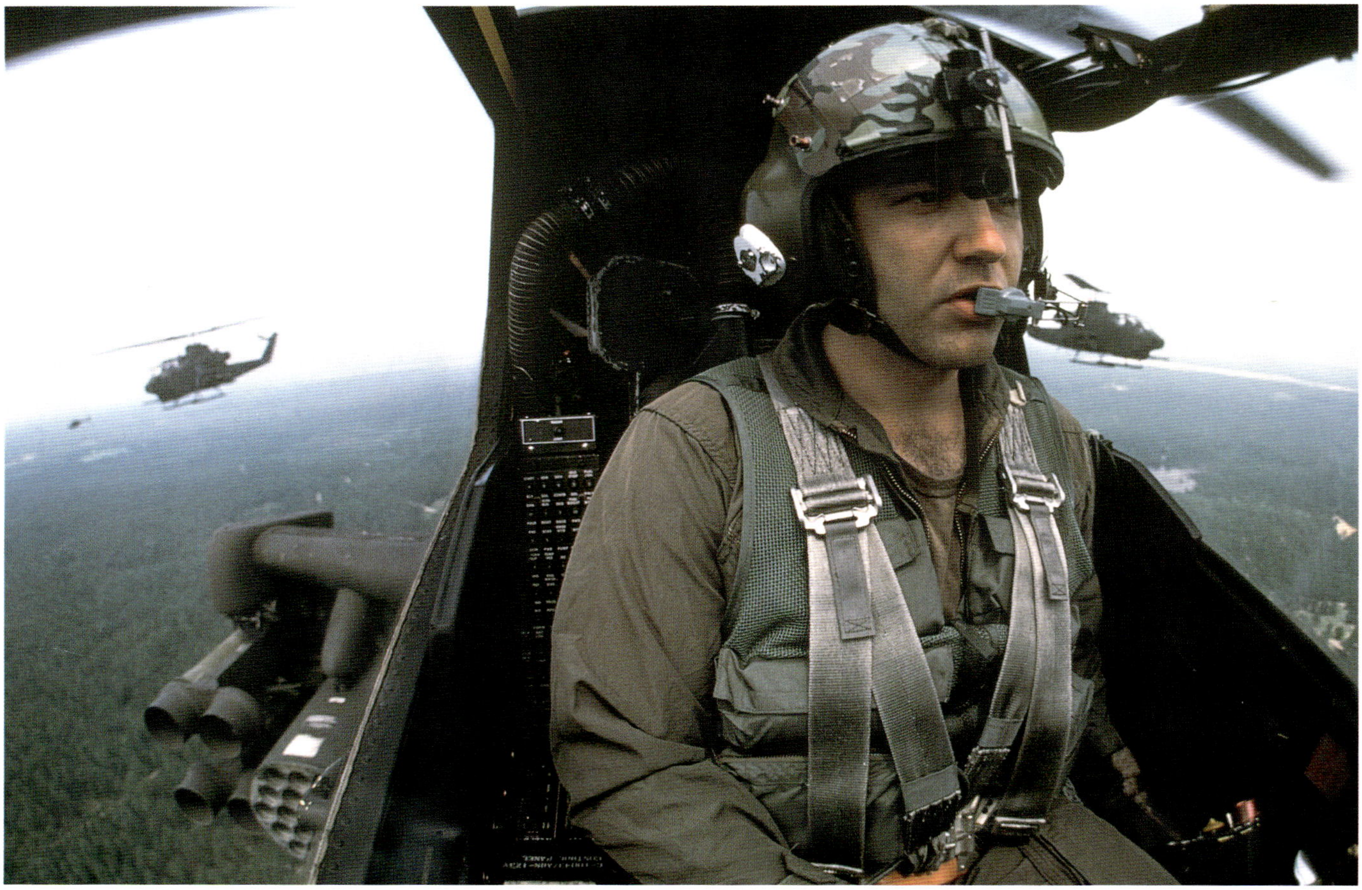

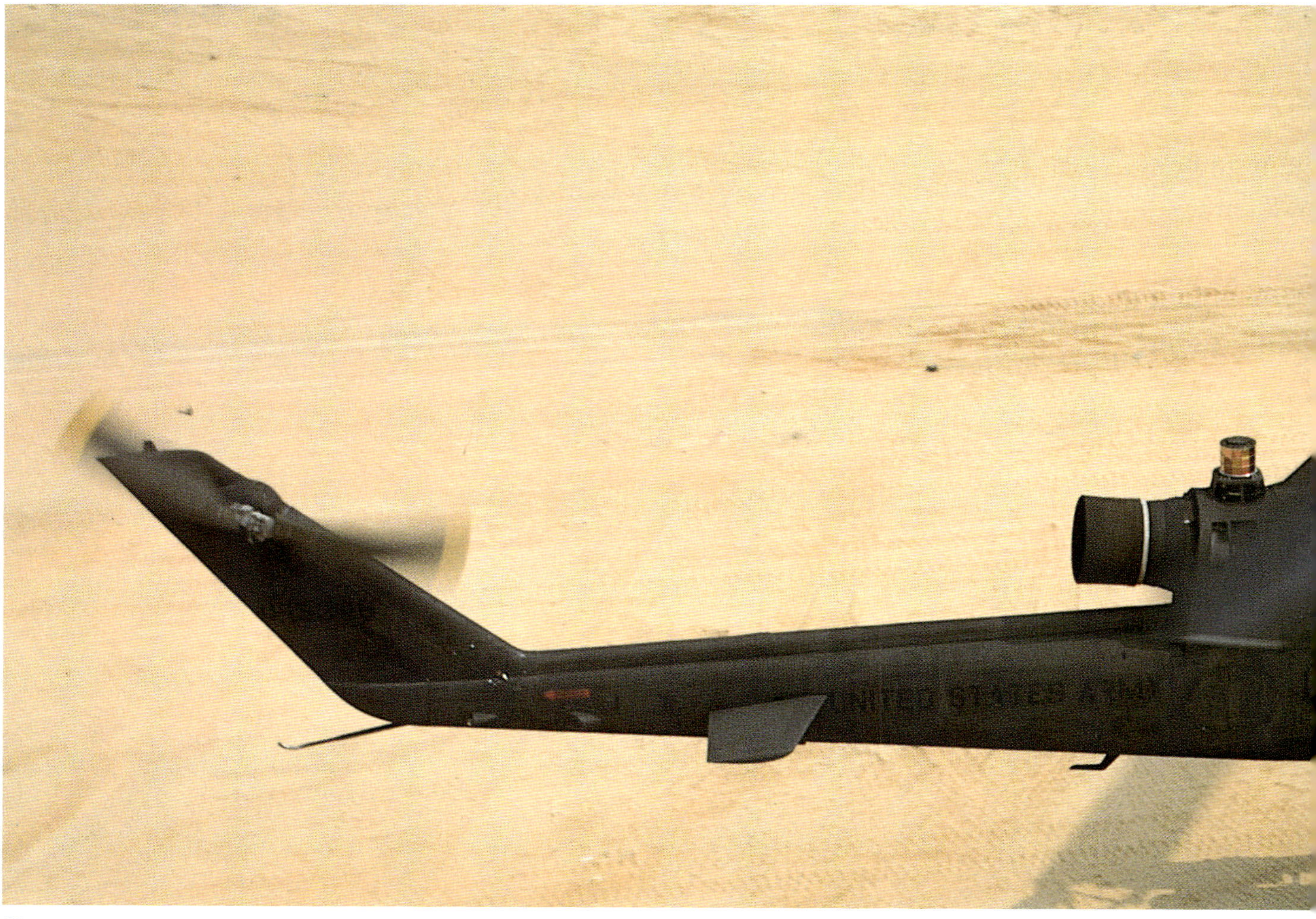

MODERNIZED AH-1S COBRA

The AH-1S Cobra, produced by Bell Helicopter Textron, is a highly efficient antiarmor helicopter. In use with the U.S. Army, the AH-1S is a model of flexibility in operations, accuracy and weapons load versatility. Equipped with four to eight TOW missiles on outboard underwing pylons, the Cobra's fire capability also includes a mixed inventory of warheads, fuses and rockets. Weapons mix can include M261 folding fin rockets (FFAR), M20 rockets, and the M197 20mm Gatling gun. The Universal Turret allows the use of 20mm or 30mm weapons.

The fire control system (FCC) features an air data subsystem, head up display (HUD), TOW sight unit, laser rangefinder, fire control computer, Doppler Navigation System and rocket management functions. Because of its system of fire interlock, the Cobra can fire its TOWs, rockets and guns at the same time without one projectile interfering with another.

The Cobra functions superbly on high-threat missions. The elimination of much canopy curvature reduces the craft's sun reflection, and the addition of an AN/APR-39 Radar Warning System gives the crew time and information to avoid enemy radar systems. Signature is further reduced by the use of a hot metal and exhaust plume suppressor, an active ALQ-144 IR jammer and special infrared signature reduction paint.

Plans for further improvements include the addition of forward looking infrared (FLIR) sensors, which will greatly enhance nighttime capability.

The result of an intensive and innovative modernization program, the AH-1S greatly increases the odds for the U.S. Army to achieve victory in engagements involving enemy tanks and other armaments.

CHINOOKS AND SEA KNIGHTS
by Geoff Sutton

TORQUE

No matter how softly it's said, the word sounds like an Anglo Saxon curse. And cursed is the helicopter pilot who has lost his anti-torque tail rotor.

Torque results when helicopter rotors are put in motion. And, unless neutralized, the inertia of the blades whirling in one direction makes the fuselage want to rotate uncontrollably in the opposite direction. It's a complex situation governed by simple physics, for, as Sir Issac Newton's Third Law of Motion states: "To every action, there is an equal and opposite reaction."

The most common approach to taming torque is the ubiquitous tail rotor with horizontal thrust sufficient to balance the twisting effects of the helicopter's main rotor, and enough thrust left over to provide yaw control.

Conventional tail rotors are the least costly to design and fabricate, but generate hidden costs by sapping the helicopter's productivity and by being dangerous to themselves and others on the ground. Tail rotors work by diverting power from main lifting rotors, which reduces payload capabilities. They are also vulnerable to ground strikes, and present a buzzsaw hazard to personnel.

A sure way to avoid the deficiencies of the tail rotor is to eliminate it. This can be done by building not one but two main rotors and stacking them one on top of the other, but geared to turn in opposite directions. These coaxial, counter-rotating blades effectively cancel torque without significant loss of lifting power; however, they offer little else in the way of additional benefits.

TORQUE FREE AND BROAD CG

Perhaps the best solution is the one developed and perfected by the Boeing Helicopter Company — the tandem rotor configuration. Tandem rotors also cancel torque by turning in opposite directions, and they offer important advantages over both conventional helicopters and counter-rotating-coaxial designs. All engine power is used directly for lifting. And since the centers of lift are widely separated, tandem rotors permit the broadest center-of-gravity (CG) range of all rotorcraft, which results in greater flexibility for handling loads.

SEA KNIGHT: FIRST TURBINE TANDEM ROTOR HELICOPTER

The H-46 Sea Knight was Boeing's first production turbine-powered helicopter. The company built over 620 of these 23,300-pound gross weight aircraft from 1960 to 1971 for both military and civilian customers in North America, Europe, and Asia.

The H-46 has a crew of three and carries up to 25 combat-equipped troops. It can take average loads of about 4,000 pounds either internally or externally. Its tandem rotor configuration results in a compact aircraft totally insensitive to wind speed and direction while hovering, landing or taking off, an attribute that makes the Sea Knight an excellent helicopter for shipboard operations.

The U.S. Navy and Marine Corps have long operated H-46s in a fleet composed of three different models — the A-model (CH-46A, HH-46A, and UH-46A), the D-model (CH-46D and UH-46D) and the E-model (CH-46E).

The Navy uses the Sea Knights primarily for vertical replenishment of ships at sea, or VERTREP, and the Marine Corps uses the aircraft for troop assault. During the early 1990s, Bell-Boeing V-22 Osprey tilt-rotor aircraft will begin to replace the Marines' H-46s, many of which will be turned over to the Navy because of their superior wind-bucking characteristics.

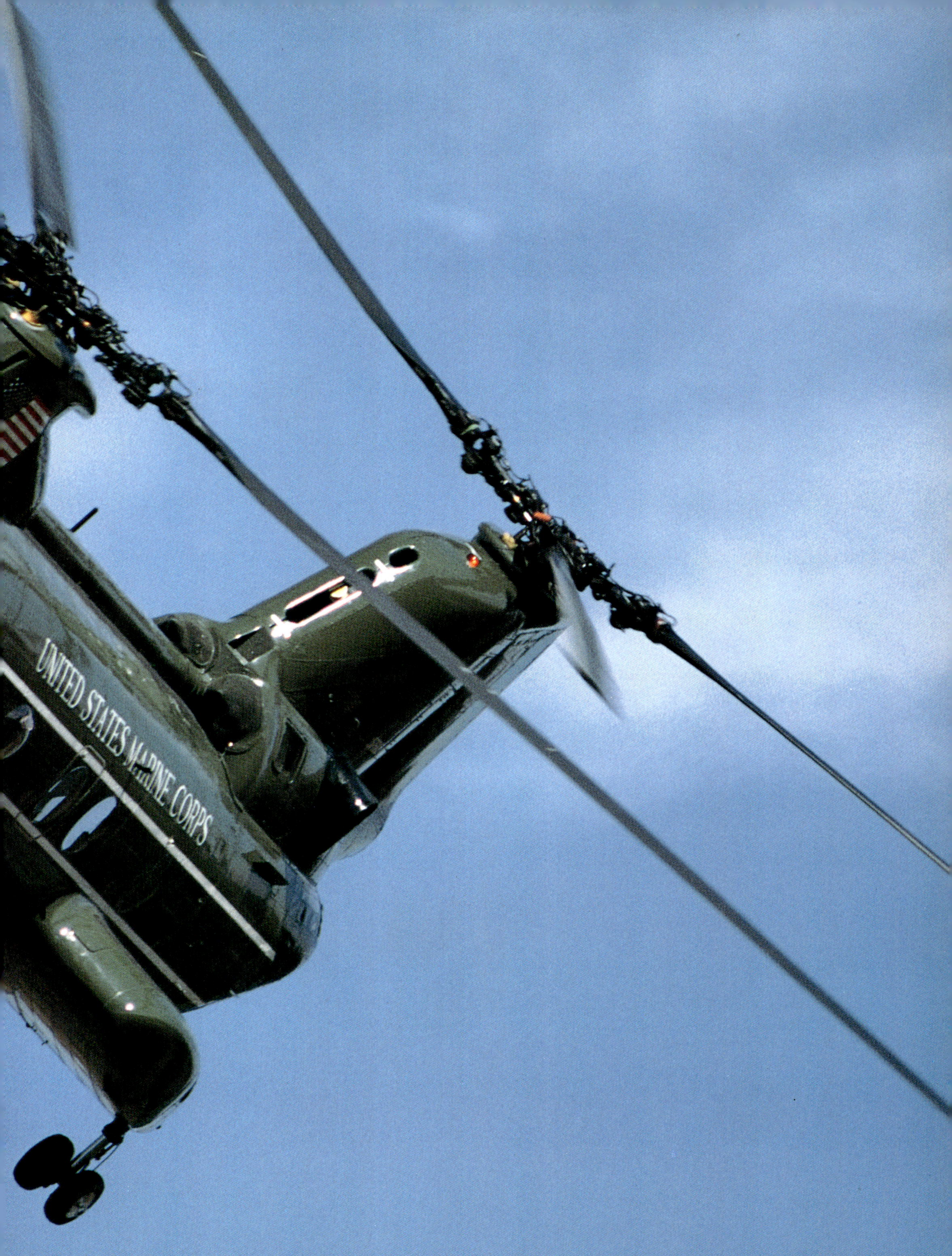
UNITED STATES MARINE CORPS

The Navy contends there is no aircraft that can perform the VERTREP mission as well as the H-46 Sea Knight and, therefore, plans to continue to operate H-46s well into the 21st century. To achieve this goal, Navy and Marine Corps aviation planners in the late 1970s developed a program that will allow the H-46 fleet to operate safely and economically past the year 2000. It involves the installation of Safety, Reliability, and Maintainability (SR&M) kits.

SR&M kits are manufactured by Boeing and are installed by Navy personnel during regular standard depot-level maintenance (SDLM) activities. (SDLM is the most extensive form of repair or modification permitted on aircraft outside the manufacturer's facilities.)

SR&M kits — it must be emphasized — are not intended to alter flying characteristics or capabilities in any way. They are designed solely to allow H-46s to continue operating efficiently and effectively through the incorporation of 26 modifications grouped into six major categories — electrical systems, hydraulic systems, avionics, rotor drive systems, airframe, and landing gear.

The first SR&M kits were delivered in July 1985, and deliveries presently continue at a rate of 10 per month. The current program schedule calls for delivery of the 357th and final kit to the Navy in January 1989.

Deliveries of SR&M-kit-equipped H-46s to the fleet are expected to be completed in mid-1989, which will eliminate all A-model designations. Since H-46A and D models are almost alike, an SR&M-modified H-46A is redesignated a D model, hence there will be only H-46Ds and H-46Es in the future.

In addition to the SR&M program, several other H-46 improvements are underway and these kits are planned for deliveries extending to 1991. Among these improvements is an increase in fuel capacity that will double the helicopter's mission radius. Other plans call for Marine Corps H-46s to acquire night-vision-capable cockpits, with some of them gaining avionics improvements that include digital navigation systems and automatic approach-to-a-hover capability.

Most H-46s will obtain provisions for the installation of emergency flotation systems intended to prevent loss of an aircraft damaged by a forced water landing. Although the H-46 has a sealed hull, it lacks water-tight compartments to make it a genuinely amphibious helicopter. Thus a broken chin window can cause serious flooding, at times enough to sink the helicopter.

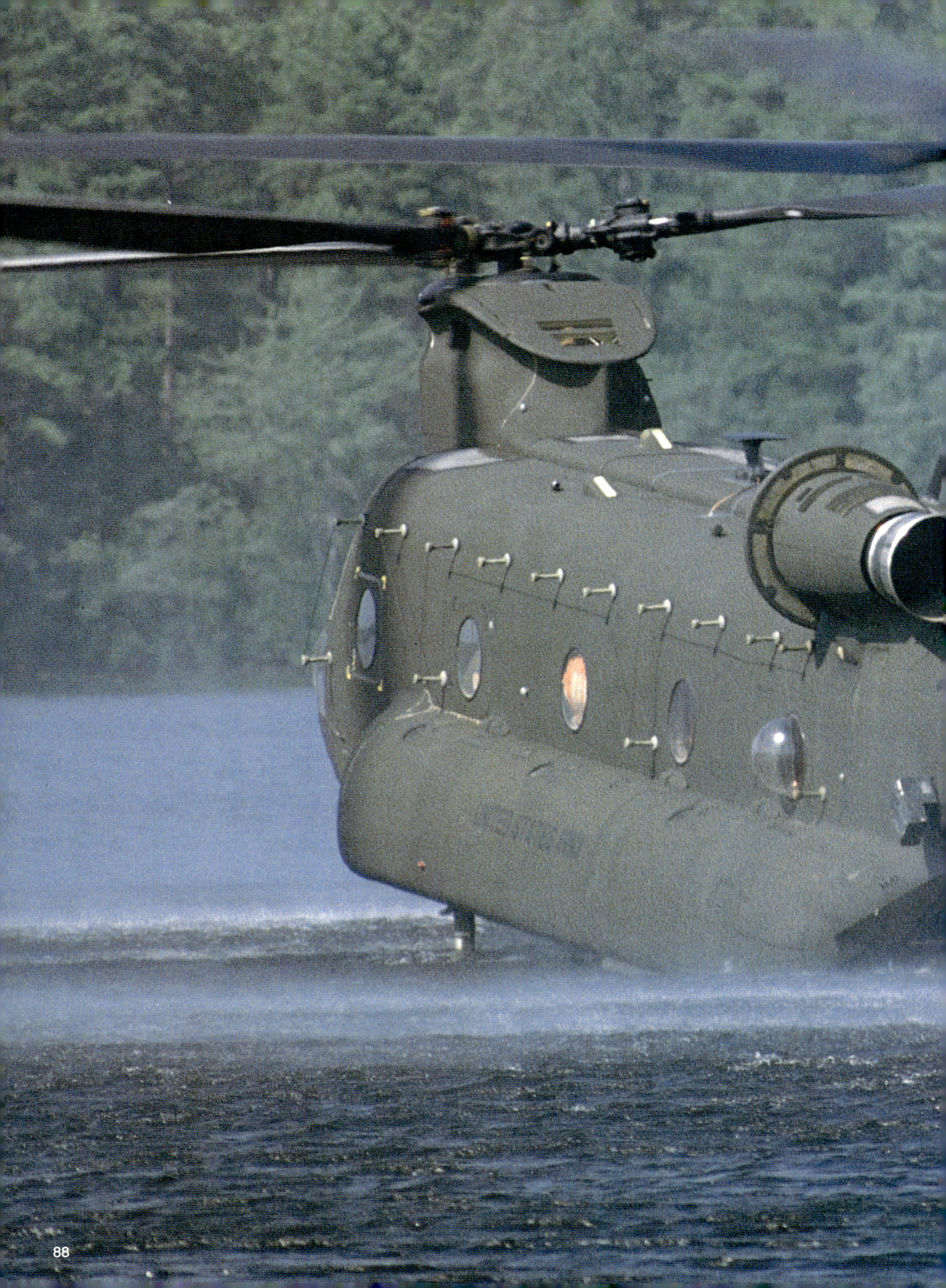

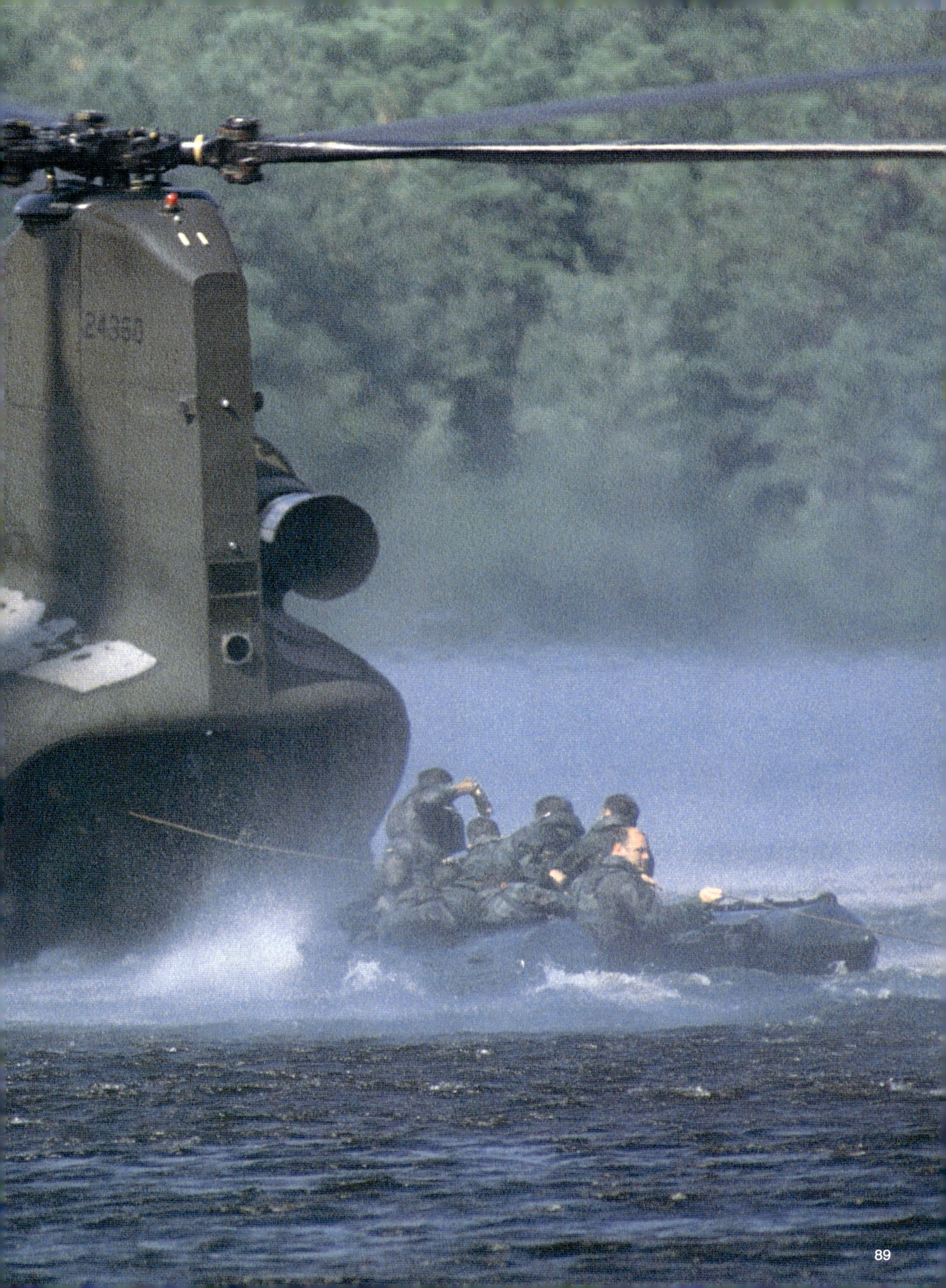
24360

WORLD'S ONLY PRODUCTION TANDEM ROTOR

"Old soldiers never die," to twist Gen. MacArthur's words, "they just fly away." Proof of that is seen in Boeing's CH-47 Chinook, which is the only tandem rotor aircraft still in production, in both the United States and Japan.

In late 1987 the Chinook celebrated its silver anniversary, for it was on Dec. 19, 1962, that the first production CH-47A Chinook was delivered to the U.S. Army. Boeing built 354 CH-47As before introducing and delivering the first CH-47B on May 10, 1967. Less than one third as many B-models were produced as were A-models — only 108. Quickly following the B-model, on March 30, 1968, came delivery of the first CH-47C, of which 270 aircraft were built.

Numerous CH-47A, -47B, and -47C Chinooks supported combat units during the Vietnam War. Chinooks have been selected by the armed forces of 16 nations, but the U.S. Army remains the largest user of the type.

The Chinook is the U.S. Army's only medium-lift helicopter. Its primary mission is the movement of artillery, ammunition, personnel, equipment, supplies and fuel on the battlefield, frequently at night and in nearly all weather conditions.

In October 1987, the Army's last remaining CH-47A Chinook was returned to Boeing's Philadelphia, Pa., production line for induction to the CH-47 modernization program. In the early fall of 1988 it will emerge as a much more powerful and productive CH-47D Chinook, which has the greatest payload of any Army helicopter — and a useful load nearly double that of the earliest CH-47A.

Currently the Army plans to upgrade 472 early model Chinooks, converting them into CH-47Ds with benefits extending beyond mere productivity. Chinook modernization is estimated to save taxpayers some $5 million per airframe — or more than $2 billion over the life of the production program — which is expected to continue through 1993.

CH-47D low-rate production began in 1980. The first 88 Chinooks to be modernized were funded by single-year contracts. In April 1985, as CH-47D production increased to four aircraft per month, the U.S. Army awarded Boeing a $1.17 billion multi-year procurement contract to rebuild an additional 240 Chinooks, saving taxpayers a further $123 million compared to the estimated total cost of five single-year contracts.

Rebuilding a Chinook requires painstaking care, which begins by removing its rotors, engines, transmissions, and other assemblies, stripping the airframe to bare skin and ribs, and inspecting every square inch. Discovery of hidden damage or corrosion is corrected by installation of new materials or components. Finally, the helicopter is reassembled to include 11 major new systems that substantially improve its operational characteristics and increase its reliability, availability, and maintainability.

Among the new CH-47D's systems are two 4,075 shp Textron Lycoming T55-L-712 turboshaft engines, improved 7,500 shp main transmissions, newly designed fiberglass rotor blades, redundant and improved electrical systems, new automatic flight control systems, and single-point pressure refueling. These and other improvements are expected to extend the helicopter's service life by 20 years or more, making it highly likely that many Chinooks can continue flying beyond their golden anniversaries.

Tandem rotor design greatly enhances the Chinooks's productivity by allowing full use of its entire fuselage; no space is lost to an empty tail boom. The cabin measures 30.2 feet long, 8.3 feet wide, and 6.5 feet high, providing a floor area of 226 square feet and a useable volume of 1,474 cubic feet, the largest of any U.S. Army helicopter. The cabin's constant cross-section is typical of most modern military transports, and results from placing the helicopter's 1,028-gallon fuel supply externally in pods on each side.

Loading and unloading a Chinook is accomplished through a full-width rear ramp, which can be lowered to the ground or raised to match the level of truck beds. One man using a built-in winch can pull cargo in or out of the aircraft. If necessary, extremely long loads are allowed to extend beyond the rear ramp (the advantage of a tandem rotor's generous CG).

The Chinook's high rotor blades permit personnel to enter or exit the helicopter quickly and safely, even when the rotors are turning. The Chinook ordinarily accommodates 33 fully equipped troops in standard sidewall seating, but can take up to 44 soldiers when additional seats are rigged in the center aisle. The cabin can also accommodate 24 standard NATO litters and two medical attendants for medical evacuation missions.

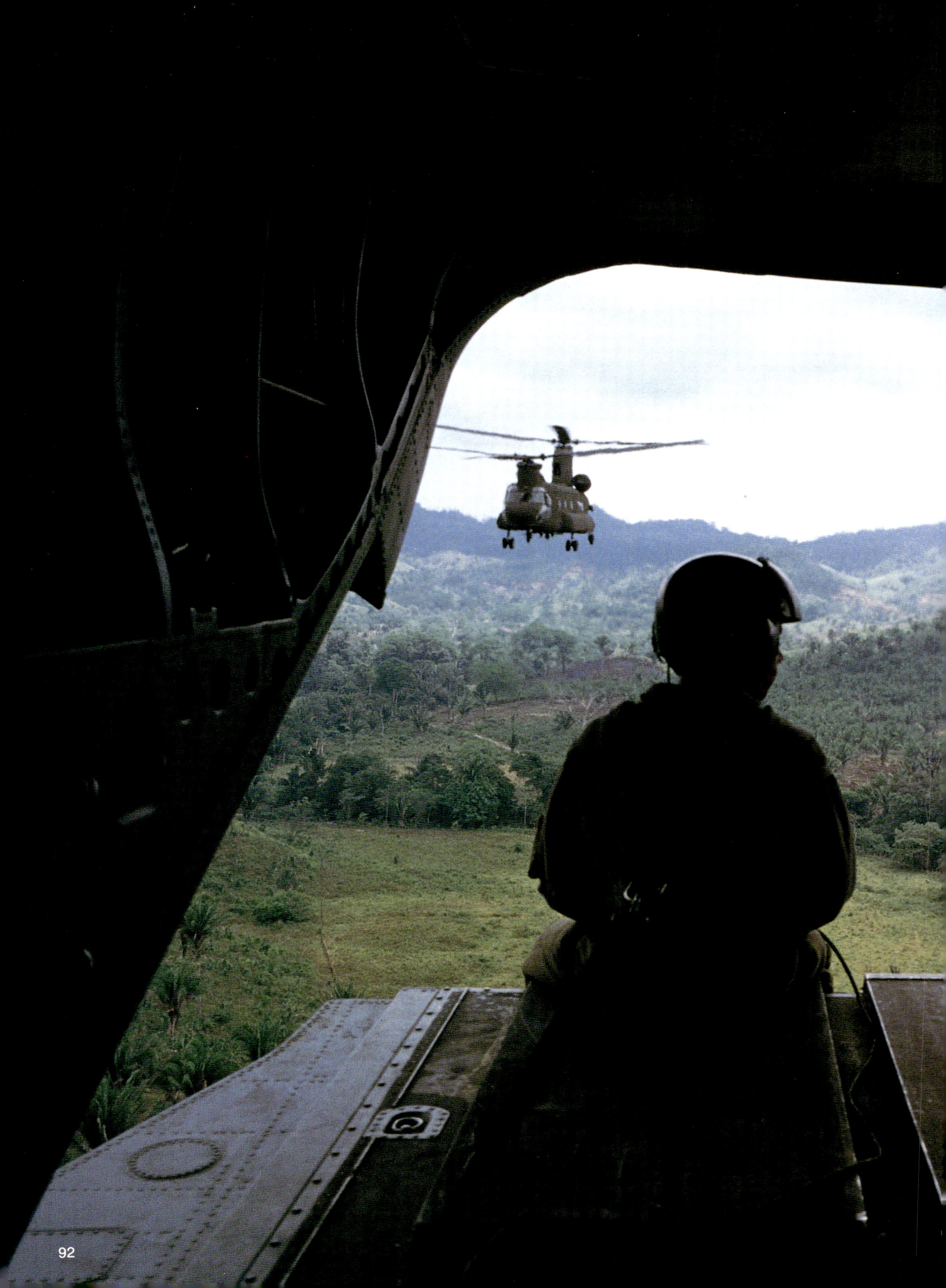

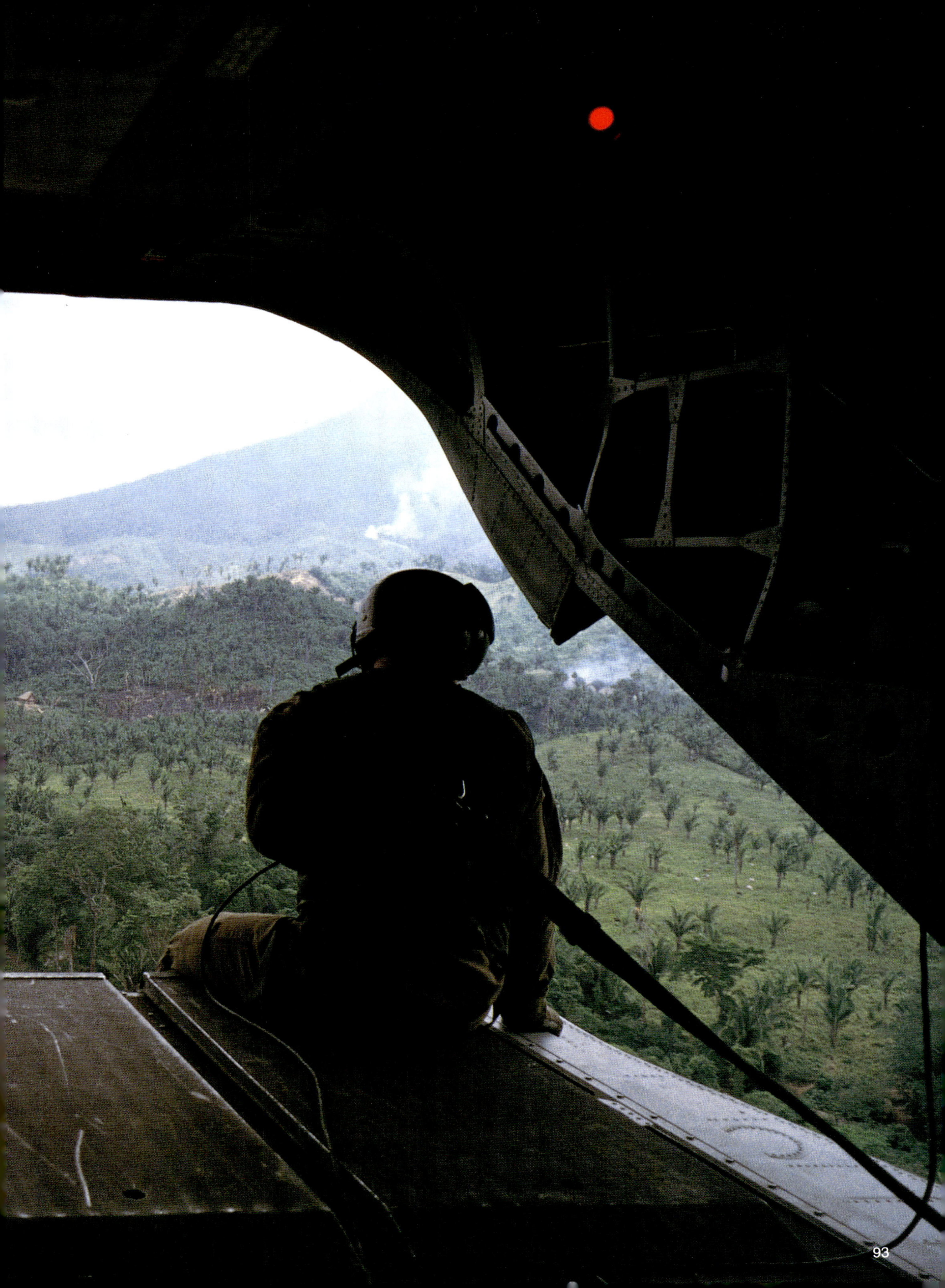

The CH-47D employs three external cargo hooks. The center hook has a capacity of 26,000 pounds; the forward and aft cargo hooks are rated at 17,000 pounds each. The helicopter's maximum gross weight is 50,000 pounds.

The Chinook's triple-hook system allows multiple suspension of external cargo, which results in greater load stability, permitting these loads to be flown at speeds up to three times faster than can be done with single-point suspension — or more than 115 knots compared to the previous limits of 40 to 60 knots.

This increased speed allows ground commanders to control much larger areas in the same unit of time. This is especially true when the CH-47D is called upon to move the M198 155mm howitzer. A single CH-47D can carry the M198, its 11-man crew, and 32 rounds of ready ammunition suspended in a cargo net beneath the gun — a payload of about 22,000 pounds.

The Chinook's three hooks also permit multiple-destination resupply missions, allowing fuel blivets, cargo containers or ammunition pallets to be emplaced at up to three separate destinations per sortie.

The combination of higher mission speeds and increased cargo capacities makes the CH-47D Chinook as effective as three U.S. Army UH-60 Black Hawk helicopters, and also gives it the lowest cost per ton-nautical-mile flown of any Army helicopter.

BUILT FOR TOUGH ENVIRONMENTS

The CH-47D is designed to function in the harshest environments — from -65 to +125 degrees Fahrenheit. Six low-pressure tires offer more than adequate flotation for most conditions, and skis can be added to keep the helicopter from sinking into snow, mud, or marshy terrain. The aircraft's unusual quadricycle landing gear provides maximum stability while loading or unloading internal cargo. Its wide stance prevents rollovers when landing at unprepared sites — including slopes up to 20 degrees.

The Chinook's hull is sealed at the factory so it can land and take off from water in conditions up to Sea State 3. If the helicopter's optional power-down ramp hydraulics have been added, and its optional water dam installed in the aft fuselage, then its rear ramp can be lowered while the aircraft is on the water. The Chinook's unique amphibious capability increases the number of available landing areas considerably.

Chinook availability is rarely limited by weather, and certainly not by darkness. The aircraft is approved for instrument operations, even in conditions of light icing and moderate turbulence. Advanced automatic flight control and stability-augmentation systems significantly reduce cockpit workload and give the helicopter flying qualities closely matching those of fixed-wing aircraft.

The Chinook's improved night-vision-goggle-compatible cockpit allows it to operate extensively and safely at night.

The CH-47D Chinook is completely self-sufficient and can operate in the field for extended periods. Its T-62T-2B auxiliary power unit furnishes all needed electrical and hydraulic power for maintenance checks and main engine starting. A panel in the aft cabin provides 26 separate monitoring and inspection functions for the aircraft's engines, transmissions and hydraulic systems.

The advanced CH-47D Chinook's systems have been modularized and simplified so that major components are more readily accessible and fewer support personnel are required to maintain them. Built-in handholds, flush steps, walkways, and integral work platforms eliminate any need for workstands and ladders.

Self-sufficiency, ruggedness, compact design, spacious cabins, the ability to operate when and where other helicopters cannot, and the ability to carry loads defying CG limitations of other aircraft — these are the hallmarks of tandem rotor helicopters. Boeing's H-46 Sea Knight and CH-47D Chinook are the epitome of the design, and both will remain in service long enough, it is said, for the grandsons (and granddaughters) of the first Sea Knight and Chinook pilots to also fly them.

CH-47 CHINOOK CHRONOLOGY

EVENT	DATE
Chinook picked as U. S. Army's medium-lift helicopter	May 1959
First flight of Chinook prototype (YHC-1B)	September 1961
First delivery of production CH-47A	Dec. 19, 1962
First delivery of production CH-47B	May 10, 1967
First delivery of production CH-47C	March 29, 1968
CH-47D full-scale engineering development contract	1976
First flight of prototype CH-47D	May 11, 1979
Production phase of CH-47D program begins	1980
First del. of CH-47D to U.S. Army	March 31, 1982
First CH-47D del. to 101st Airborne Div., Ft. Campbell, Ky.	Feb. 28, 1983
CH-47D initial operational capability (101st Abn. Div.)	Feb. 28, 1984
CH-47D production rate fixed at four units per month	April 1985
Boeing awarded $1.17 billion CH-47D MYP contract	April 8, 1985
Del. of 2,500th CH-47 fiberglass	May 16, 1985
First CH-47D del., 24th Infantry Div., Ft. Stewart, Ga	Aug. 29, 1985
First CH-47D del., 82nd Airborne Div., Ft. Bragg, N.C	Jan. 16, 1986
First CH-47D del., 9th Inf. Div., Ft. Lewis, Wash	July 24, 1986
First CH-47D del., III Mobile Armored Corps, Ft. Sill, Okla.	Feb. 6, 1987
First CH-47D del., 4th Inf. Div. (Mech), Ft. Carson, Colo.	April 28, 1987
First CH-47D del., 6th Cav. (Air Combat), Ft. Hood, Texas	July 14, 1987
CH-47D del. completed to active Army in continental U.S.	mid-1987
First CH-47D del., 205th Avn. Co. Finthen AAF, Germany	Oct. 30, 1987
Planned completion, CH-47D del., U.S. Army in Germany	late-1988
Planned completion, del. of 472 CH-47DS to U.S. Army	late-1993

MODEL 107/H-46 SEA KNIGHT CHRONOLOGY

EVENT	DATE
First flight of Model 107 (H-46 predecessor)	April 1958
CH-46A contract award from U.S. Navy	September 1961
First CH-46A rollout	April 1962
First del. of CH-46A to U.S. Marine Corps	May 1962
First del. of UH-46 to U.S. Navy for VERTREP	June 1964
SR&M modification kit program begins	Dec. 20, 1980
First flight of SR&M-equipped CH-46E	Nov. 21, 1983
Del. of first production SR&M kit	July 31, 1985
Del. of first SR&M-equipped CH-46 to fleet	December 1985
Planned completion of SR&M program	January 1989

95

Boeing CH-46